Above and Below

MODERN PHYSICS FOR EVERYONE

Jack Challoner

Published by Explaining Science Publishing.

ISBN: 978-0-9954750-3-8

DEDICATION

This one is for my father (1912 - 1998),
who inspired me to be curious.

CONTENTS

ACKNOWLEDGEMENT

Thank you to the scientists, alive and dead, who have come up with the theories and discoveries explained in this book. You didn't have to, but you did.

INTRODUCTION

This book is for anyone who is interested in understanding something of the amazing progress science has made in figuring out the big questions about the Universe in which we live – from the smallest particles to the largest galaxies. It contains no equations – well, okay, just one – and is written for adult readers with no scientific background. That's not to say there aren't challenging parts, which you may have to read over twice, or just 'get the jist' of. If there are parts that bamboozle you, please just read on – I hope that you can get a flavour of the kinds of spellbinding theories science has come up without having to understand all their details.

Occasionally, you will find numbers in 'scientific notation'. This is a way of writing big and small numbers powers of 10. The number raised to the fifth power, for example, is written 10^5, and is equivalent to 1 with five zeroes after it – in other words, 100,000. The number 234,000 is then written 2.34×10^5. Similarly, powers can have negative values: 10^{-5}, for example, is a 1 five places after decimal point – in other words, 0.00001. In this case, 2.43×10^{-5} is equal to 0.0000243. You can imagine how useful scientific notation can be when writing some of the extremely small and extremely large numbers that describe our Universe.

One more thing: Chapter One is an overview of our place in the Universe, and provides an introduction to stars, galaxies and the expansion of the Universe. If you already have a basic knowledge of these things, do feel free to skip that chapter and go straight to Chapter Two, which examines the nature of time and space.

I hope you enjoy my book, and learn something new about the Universe and science's current understanding of it. And I hope you are inspired to search online for more information about anything that interests you, primed as I hope you will be by the introduction to these fascinating topics provided here.

Jack Challoner, 2017, Bristol, UK.

1: A CENSUS OF THE UNIVERSE

We human beings yearn to make sense of our existence, and of the reality of which we seem to be part. This yearning motivates some of us to take part in the systematic journey towards truth that is 'science'. Any serious scientific attempt to explain our existence must take into account not only our immediate physical environment – the properties of matter that we can probe directly – but the entire Universe of which we are but a tiny part. I can get a good idea of what the Universe looks like on a grand scale whenever I look up at the sky on a clear, dark night. For when I look up, I am looking out into the Universe, across vast distances. Gazing into space raises a number of timeless and fundamental questions: What is all that stuff up there? Has the Universe existed forever? Is the Universe infinitely large? Astronomers have gathered so much information, and astrophysicists and cosmologists have developed such compelling and well-tested theories, that we have begun to produce answers to some of these deep questions, which human minds have pondered for thousands of years. Thanks to the astronomers' observations, we now know a great deal about the objects that populate the Universe. It is as if astronomers are in the process of conducting a detailed cosmic census. And so, in preparation for the mind-blowing concepts explained in forthcoming chapters of this book, in this first chapter, we'll take a look at this 'census'.

Just as human census data provide a snapshot of a population, from which social scientists attempt to explain that population's trends and behaviours, our cosmic census will provide a snapshot of the Universe as we see it today. It is the job of cosmologists to determine its nature, its origins and its destiny.

A cosmic overview Wherever I look in the night sky, I see points and patches of light punctuating a dark background. Many ancient astronomers believed the sky to be the inside surface of a hollow sphere with Earth at its centre; some considered the points of light to be holes in the sphere, through which light could shine from heaven. This view was shaken (to destruction) by astronomers' careful observations and mathematical theories over hundreds of years. By the mid-19th century, when astronomers began to find out what the stars are made of and how far away they are, the naive ancient ideas about the night sky were already long gone. The blackness of the night sky is a deep, three-dimensional blackness, not a black two-dimensional surface peppered with holes.

One of the most prominent features of the night sky is that it appears fairly isotropic: it looks much the same in all directions. There are no regions that are totally black, devoid of stars; there are no areas that are completely light. Everywhere, there is a seemingly random mixture of brighter stars and fainter ones. In fact, there seem to be many more fainter stars than brighter ones. This fact, together with the sky's isotropy, suggests that the Universe is probably homogeneous: that it has a fairly uniform structure and composition throughout (think homogenised milk, in which the fat globules are so tiny that they mix in and do not separate from the rest of the mixture, so that the milk has the same consistency throughout). In a homogeneous Universe, the night sky would indeed appear much the same from wherever you are, and look the same in every direction. Of course, the Universe is not completely homogeneous: if it were, it would have the same density throughout, so there could be no clumps of matter – including your body, the Earth and the stars. However, I shall assume for now that the Universe is indeed fairly homogeneous on a large scale. In the same way, a sponge has holes in it (and indeed the milk has tiny fat globules), but is nevertheless uniform as a whole.

The homogeneity of the Universe can help me answer two of the questions that I posed above: "Has the Universe existed forever?" and "Is the Universe infinitely large?" The answers to these questions cannot both be "yes": in an infinitely-large, homogeneous Universe, my line-of-sight in every direction would eventually end at the surface of a star, and the sky would therefore appear bright, not dark. Of course, the stars in some directions could be so far away that their light has not reached us yet – in which case, the Universe may be infinitely large, but it cannot have existed forever. The conclusion that the Universe cannot be both infinitely large and infinitely old is disturbing, and throws up further questions, such as "If the Universe is not infinitely old, what was there before the Universe?" and "If the Universe is not infinite in space, then what is outside it?" This line of reasoning is called Olbers' Paradox, after the German astronomer Heinrich Olbers (1758 - 1840), who popularised it in the 19th century. (In case you are thinking there might be objects in the way blocking the light from some of the distant stars ... well, those objects would absorb radiation from those stars, and over an infinite amount of time, would heat up to temperatures that would make them glow as bright as the stars they block.)

On closer inspection of the night sky, it seems not to be quite as isotropic as I suggested: a certain region of the sky is richer in stars than others. And, in a sufficiently clear, dark sky, or in photographs taken with long time exposures, a broad, fuzzy, milky-white band is seen across that same part of the sky. Close inspection of this band, called the Milky Way, reveals the fact that it is actually made up of countless faint stars. When we look at the Milky Way, we are gazing into our own cosmic city, the Milky Way Galaxy, from our position in the outskirts.

Star city The Milky Way Galaxy is a huge collection of stars and other material so vast that it takes light, travelling 300,000 kilometres every second, about 100,000 years to travel its full extent. All of the 6,000 or so stars we can see with the naked eye are inhabitants of this cosmic conurbation – which, according to recent estimates, contains around 300 billion (3×10^{11}) stars altogether. All of the stars of the galaxy are associated with one another by the invisible, attractive force of gravity – the same force that keeps Earth in orbit around the Sun. If there were no gravitational force between the stars of the Milky Way Galaxy, they would have moved off into deep space long ago; there would be no galaxy. In fact, without gravity, there would be no stars, for gravity is responsible for their formation.

Stars form from the interstellar medium: gas and dust that permeates much of the space within the galaxy. Interstellar gas contributes just 2% of the galaxy's mass, the dust a tiny fraction of 1%. Most of the space between the stars is practically empty: typically, there is an average of one particle per cubic centimetre. The interstellar gas is made of hydrogen atoms (about 75%), helium atoms (about 25%) and traces of a few other elements, notably carbon, silicon and oxygen. The interstellar dust consists of grains much smaller than dust we find in our homes: they are closer to the size of smoke particles. They are typically about 1 micron (one millionth of a metre) in diameter and consist of a few thousand atoms. In most of interstellar space, there is one dust particle for every trillion (10^{12}) hydrogen atoms.

In some regions of interstellar space, however, there are about ten thousand particles in every cubic centimetre: much more dense than most of interstellar space, but still as rarefied as the best vacuums produced in laboratories on Earth. These regions are called giant molecular clouds, because they are cold enough and dense enough for the hydrogen within them to exist as molecules, rather than as individual atoms. Giant molecular clouds, and smaller versions called Bok Globules, often appear as eerie dark patches in the night sky because the dust they contain obscures the stars behind them. Since these regions are denser than the rest of the matter in interstellar space, self-gravity – the mutual attraction of all the matter present – can cause them to collapse in on themselves. This process is the beginning of star birth.

A star is born The gravitational energy released by the collapse of a molecular cloud causes the gas and dust to heat up, and the temperature eventually becomes high enough to initiate nuclear fusion, the process in which small atomic nuclei fuse together to form heavier ones. The predominant nuclear reaction inside stars is the fusion of hydrogen nuclei to make helium nuclei. Fusion reactions release huge amounts of energy, which heats the star even more. The heat increases the pressure of the gas, which supports the young star, dubbed a protostar, against further gravitational collapse. This stable configuration, in which gravitational collapse is balanced by internal pressure, is called the main sequence. Ordinary stars in the main sequence are called dwarfs – a rather misleading term designed to differentiate them from giant stars, which dwarf stars become towards the end of their lives.

Heat passes from the core of the star outwards to the surface. The core of a main sequence star like the Sun has a temperature of several million degrees Celsius, its surface a temperature of several thousand. Like any hot object, a star's surface gives out a wide spectrum of electromagnetic radiation: radio waves, infrared, visible light, ultraviolet. A layer of gas surrounding the star, called the corona, is heated to a temperature high enough to emit X-rays and, in some cases, even gamma rays. Radiation emitted by a protostar exerts pressure, which pushes much of the remaining dust and gas away to reveal a naked, fully-formed star.

Astronomers can deduce a great deal about a star from the radiation they receive from it. A graph showing the intensity of each wavelength of radiation, for example, provides a direct indication of the star's surface temperature; we can tell how hot a star is without being anywhere near it. We can also tell what chemical elements are present in a star by studying dark lines in the spectrum of the light coming from the star. (The lines that appear in the spectra of stars correspond to the lines observed in laboratory-based spectroscopic analysis of the elements present here on Earth. Incidentally, this, along with a host of other pieces of evidence, strongly suggests that the matter 'up there' is the same as the matter 'down here'. So strong is this conviction that, when an unrecognised line was identified in the spectrum of the Sun in 1868, it was assumed to be an element that does exist on Earth, but that had simply not yet been discovered. This indeed turned out to the be the case: the element was helium, named after the Greek word for the Sun, *helios*.)

The fact that matter in space is made of the same ingredients as matter here on Earth means that it behaves according to familiar physical laws, albeit often in extreme circumstances. This enables astrophysicists to make predictions of objects so far undetected, on the basis of current theories of matter, which can guide astronomers in what to look for next. In return, astronomers' observations – or lack of observations – of the predicted objects reflect on the quality of the theories. A good example of this is the discovery of brown dwarfs. The first brown dwarf stars were detected by infrared telescopes in 1995, but they had been predicted by theories developed 30 years earlier. A brown dwarf is a 'failed star', born from a small protostar that never quite made the main sequence, because its core never reached the temperature necessary for fusion to take place. Because of their size and the fact that they emit little or no visible light, brown dwarfs are very difficult to detect. Today, however, many examples of brown dwarfs are known, and astronomers now believe that they are as numerous in our galaxy as main sequence stars.

Nebulas and planets Ultraviolet radiation emitted by a hot young star causes hydrogen atoms in any nearby interstellar gas to emit a characteristic glow – in much the same way as fluorescent lamps do. The gas may then be visible from Earth as an emission nebula: a stunning red 'cloud' in space. The red light is characteristic of hydrogen; astronomers refer to it as 'hydrogen-alpha'. Interstellar dust can also be visible in the night sky, as reflection nebulas: blue glowing clouds. The blue glow is produced when visible light from stars - a mixture of all the colours of the spectrum, from red to blue - hits the dust grains and is scattered by them (reflected in all directions). Because of the size of the dust particles, light from the blue end of the spectrum is scattered more than light from the red end. This mechanism is called preferential Rayleigh scattering, and it also explains why our daylight sky is blue, although in that case, sunlight scatters off air molecules, not interstellar dust. The dust also causes a star to appear dimmer overall than it would if space were dust-free – a phenomenon called extinction, which can make it difficult for astronomers to determine how far away the stars are. Extinction by interstellar dust can even block the light completely from very distant stars, or stars that are behind a dense cloud.

Once a protostar has condensed to make a new star, and pushed away most of the gas and dust leftover from its creation, some material may still remain. The protostar and its surrounding gas would have been rotating, so this material forms into a spinning disk, and may then clump together at various points because of its self-gravity, to make not stars, but planets. Several hundred stars other than the Sun are already known to have attendant planets – a figure that is growing all the time. It seems reasonable to suppose that there are billions of planets in the Milky Way Galaxy. It is also reasonable to suppose that some other planets harbour some kind of life.

Our own Solar System may, then, be fairly typical. In addition to the known planets, there are huge numbers of small rocky bodies called meteoroids, and ice- and dust-covered rocks that are found in a huge spherical cloud that surrounds the entire Solar System, called the Oort Cloud. The objects found in the Oort Cloud are, like the planets, made of material left over from the molecular cloud from which the Solar System formed. But unlike the planets, they had enough speed to escape from the central part of the system. Occasionally, they 'fall' in towards the centre of the Solar System, acquiring highly elliptical orbits that take them very close to the Sun. When one of these objects nears the Sun, some of its gas and dust is released as it is heated by solar radiation, and it may then be visible from Earth as a long-tailed, fuzzy, and in some cases spectacular object: a comet.

Star death Most stars remain in a stable main sequence configuration for a few billion years. The temperature of a star's surface depends upon the star's size: larger stars are hotter, and appear bluer; smaller stars are cooler and appear yellow or red. About 90% of the stars in the Milky Way Galaxy are red or yellow dwarf stars, less than 1% are hot blue stars. This is because the larger and hotter a star is, the faster it uses up its supply of hydrogen, and the shorter its life will be. But the lives of all stars eventually come to an end – a process that begins when their supply of hydrogen 'fuel' for fusion dwindles.

Our Sun, a yellow dwarf, has shone for around 5 billion years, and astrophysicists estimate that will remain a main sequence star until 5 billion years from now. By then, most of its hydrogen will have been fused to make helium, and nuclear fusion will slow down to such a degree that gravitational collapse will dominate once again. The Sun will begin to shrink, and this will heat the Sun's core to even greater temperature – enough to initiate a new round of fusion. This time, helium nuclei fuse to form nuclei of heavier elements including carbon and oxygen. Meanwhile, the increased temperature of the Sun will cause the outer layers of gas to swell, encompassing the orbit of the inner planets, including Earth. At this point, the Sun will be a red giant. About 0.5% of stars in the Milky Way Galaxy are currently in this red giant phase. After a few million years as a red giant, the supply of helium will run out, and the Sun will once again begin to collapse under gravity. Eventually, another stable state will be reached: the Sun will become a 'white dwarf' about the size of the Earth. About 9% of the stars in our galaxy are white dwarfs. At this stage, the gases surrounding the core will be pushed away into space, forming a vast ring called a planetary nebula. When stars significantly larger than the Sun come to the end of their main sequence stage, they become not red giants, but red supergiants. Being much more energetic, they can sustain fusion to the point where iron is produced at their cores. But at this point even they stop fusing, and gravitational collapse takes over once again. They flare up dramatically as supernovas. On average, there is one supernova somewhere in the galaxy every 30 years.

Whether a star goes supernova or becomes a white dwarf, it throws off some of the material in its outer layers – a mixture of hydrogen, helium and the heavier elements that were produced by fusion during its main sequence – far into space. These elements become part of the interstellar medium, and may be incorporated into new stars. New stars born from this enriched, recycled interstellar medium are called Population I stars. In fact, most of the galaxy's stars are Population I stars, including the Sun. They are often referred to as 'metal rich' (in this context, astronomers generally refer to any elements heavier than hydrogen and helium as metals). It is likely that rocky planets like Earth will only be found around metal-rich, Population I stars, since elements heavier than helium are needed to make rocks. Older stars – which formed when the galaxy was young and consisted mostly of hydrogen and helium – are called Population II stars. These stars are referred to as 'metal poor'. However, even Population II stars are not 'first generation': they, too, contain some 'metals'. The very earliest stars are Population III stars, which should in theory consist only of hydrogen and helium, with a tiny amount of lithium. No Population III stars have yet been detected; most will no doubt have died long ago.

Strange ends The ultimate destiny of a star depends upon its initial mass. As we have seen, the remnant of a star with a mass similar to the Sun is a small, dense, stable object called a white dwarf. A teaspoonful of white dwarf material would have a mass of several thousand tonnes. What prevents the further collapse is an increase in the internal pressure of the star caused by a strange phenomenon called electron degeneracy, which will be explained in Chapter Four. Stars that are considerably more massive than the Sun do not end up as white dwarfs; their crushing self-gravity is strong enough to overcome electron degeneracy, and they collapse to an even more dense state called a neutron star. One teaspoonful of the material from the centre of a neutron star would have a mass of hundreds of millions of tonnes. The gravitational forces are so strong inside the collapsing star that protons and electrons are forced to merge, one proton and one electron making one neutron. What stops the collapse of these stars any further is a phenomenon called neutron degeneracy; again, this strange effect will be explained in Chapter Four.

Neutron stars spin; all stars do. But, like a skater who starts spinning slowly then speeds up as they bring their arms in close to their body, tiny neutron stars spin much faster than the much larger stars from which they formed. Typically, neutron stars spin several times per second – very fast compared with once every 30 days or so for the Sun. The rotation creates electric currents inside the star, which in turn produce magnetic forces. Free electrons still exist around the star, and the magnetic forces drag these electrons around at speed, creating a beam of radio waves at the top and bottom of the star. Astronomers can measure the spin of a neutron star because of these radio beams, which can be detected by radio telescopes here on Earth. Spinning neutron stars are called pulsars, because the now-you-see-it-now-you-don't effect of the rotating beam of radio waves appears as a succession of regular pulses of radio energy, just as a beam of light from a lighthouse in fog appears as a regular pulsing of light. A pulsar's rate of spin decreases gradually over thousands of years as the pulsar loses energy as radio waves.

Because they are neutron stars – and, therefore, the remnants of massive stars – all pulsars are born in supernovae. The term 'supernova' is derived from the Latin word nova, meaning 'new': a supernova involves a many-fold brightening of a star, so that a very distant star previously too dim to be seen may suddenly appear as a new star that shines for several months, then fades. Many such 'new stars' have been recorded in history, and pulsars are observed at some of the sites at which they were seen. The classic example is the Crab Pulsar, which is found in the precise spot in the sky where a supernova had been observed in 1054. Astronomers had assumed that a pulsar exists at the heart of every supernova remnant – an assumption that was at odds with observations, since pulsars were observed only in about 1% of supernova remnants. This kind of mystery is clearly important in any census of the Universe. A partial solution came in 1998, with the discovery of a type of pulsar called a magnetar: a neutron star whose magnetism is about 1,000 times as strong as that of an ordinary neutron star, and more than one thousand million million times as strong as the Earth's. The incredibly strong magnetic forces cause the magnetar to slow down its rate of rotation much more quickly than an ordinary pulsar, and also causes sporadic bursts of energy that can be detected as very intense gamma rays. These two mechanisms result in the magnetar losing energy over a period of tens of thousands of years rather than many millions as for normal pulsars. So, at the centre of many supernova remnants may be long dead magnetars: cold dead neutron stars rotating only very slowly and producing only weak radio waves.

The existence of white dwarfs and neutron stars was predicted by applying well-tested theories of matter, including ones that describe the degeneracy of electrons and neutrons, to the high pressure conditions inside a dense, dying star. Enough observations have been made of objects in the real Universe with the predicted properties to believe strongly that they exist. Of crucial importance are measurements of the size, temperature and mass of candidate objects. What do the same theories predict might be the destiny of a star with even more mass than one that becomes a neutron star? What happens if the density and pressure of the dying star increase still further? Can the gravitational collapse of such an object overcome even neutron degeneracy? The answer, according to the best theories of matter, and again supported strongly by observations, seems to be 'yes'. Theory predicts that when gravity overcomes neutron degeneracy, neutrons – each less than one millionth of a millionth of a centimetre in diameter – break down into their even smaller, constituent parts, called quarks. If this is correct, the core of a suitably compact and massive object should become populated with densely packed, pure quarks. Astronomers have found several candidate objects whose size and mass fit those of these hypothetical 'quark stars'. I shall investigate the nature and behaviour of quarks in detail in Chapter Four.

One can imagine a star with an even greater initial mass, in which the gravitational collapse would be even more extreme. What would happen to such a star if even the quarks are crushed? Can matter withstand such forces? Theory says 'no': matter collapses in such circumstances to a point of infinite density called a singularity. And, according to the best theory of gravity, the General Theory of Relativity, the gravitational forces in the space around a singularity are so extreme that even light cannot escape, so such a region of space has been dubbed a 'black hole'. As we shall see in Chapter Three, black holes are very strange objects indeed.

It is not only light, but any type of electromagnetic radiation that is unable to escape the gravitational pull of black hole. This makes black holes very difficult to detect: they cannot emit or reflect any radiation, and are therefore completely dark. However, the existence of a black hole can be inferred from the behaviour of matter that interacts with it. In many cases, for example, astronomers have observed a star in orbit around an invisible object. From measurements of the speed of the star and the radius of the orbit, astronomers can work out the mass of the invisible object. The gravitational influence is strong enough to pull gas from the companion star and 'into' the black hole. As it falls towards the black hole, the gas heats up, forming a ring around the black hole, called an accretion disc, which is so hot that it emits X-rays. Enough observations of stars in orbit around massive, invisible objects and of hot accretion discs have been made to convince astronomers that black holes really do exist. However, there seems to be little consensus about how many black holes are lurking in our galaxy. Some astronomers believe that black holes may be very common indeed, and that there may even be countless 'primordial' black holes that formed when the Universe was very young.

The shape of our galaxy So far in our census, we have looked at the kind of objects found within the galaxy, and the nature of the space between them. But what can we say about the galaxy as a whole?

The first astronomers to attempt to work out the shape of our galaxy were the English brother-and-sister team William (1738 - 1822) and Caroline Herschel (1750 - 1848), who produced a roughly symmetrical, oval-shaped map of our galaxy, with the Sun near to the centre. Dutch astronomer Jacobus Kapteyn (1851 - 1922) formulated a similar, but slightly more accurate, model of the galaxy in a painstaking effort spanning the 1910s. Both the Herschels and Kapteyn worked out their models by counting the number of stars in small regions of the sky in all directions, and measuring their brightness. Their observations were affected by the presence of the dust – particularly in the direction of the galactic centre, since there is more interstellar space in that direction.

American astronomer Harlow Shapley (1885 - 1972) produced a much better model of our galaxy in 1920. Instead of simply counting stars, Shapley had studied the distribution of large, spherical groups of stars called globular clusters. In particular, he was looking for stars within the clusters whose brightness, or magnitude, varies over several days. The exact period of variation of the type of star he was studying (called RR Lyrae stars) is determined by the star's luminosity – the total amount of light the stars produce. And so, by studying the way the magnitude of a star varies, he could work out their luminosity. By comparing the luminosity of these stars with their average magnitude as seen from Earth, Shapley was able to estimate how far away they are, and therefore the distances to the globular clusters. He found that the galaxy's globular clusters seem to be symmetrically distributed around a point lying in the constellation of Sagittarius, and so he proposed that the centre of the galaxy lay in that direction, some 100,000 light years away from us. He also worked out that the galaxy must be disc-shaped, with a spherical halo centred at the galactic centre that extends above and below the disc. Shapley was right about the direction of the galactic centre, and the existence of the halo, but even his measurements of distance were affected by the extinction of starlight caused by the interstellar dust, of whose existence he was unaware. In fact, the centre of the galaxy lies 'only' about 26,000 light years from us.

The interstellar dust was discovered in 1930; in 1944, Dutch astronomer Hendrik van de Hulst (1918 - 2000) found a way to 'see through it'. Van de Hulst reasoned, from his knowledge of the behaviour of electrons in hydrogen atoms here on Earth, that cold hydrogen in the interstellar gas should emit radio waves with a very specific wavelength of 21.1cm. Radio waves are not scattered by the dust particles in the way light is, so they pass through interstellar space largely unimpeded. The 21cm hydrogen line, as astronomers call this radiation, was first observed in 1951. It meant that astronomers could determine the distribution of gas in most of the galaxy and, from this, work out the galaxy's shape and size. When they did, they found evidence that the galaxy has several curved 'arms' distributed around a central bulge. By 1958, they were in little doubt that the galaxy, in terms of the distribution of hydrogen in interstellar space at least, was spiral in shape. Astronomers were not surprised by the spiral shape ... but more of that later. The accuracy of the map of our galaxy has improved since the 1960s: by 1995, enough evidence had been amassed to make them conclude confidently that the galaxy is a logarithmic spiral with four main arms.

The Milky Way Galaxy is rotating – a fact that can also be deduced from the motions of the stars. Astronomers can work out how fast the Sun is travelling around the galaxy by measuring the apparent shift of the centre of the galaxy, a powerful source of radio waves, as seen from our moving Solar System. It turns out that the Sun (and the whole Solar System) is travelling at a speed of about 250 kilometres per second in a nearly circular orbit. At this speed, it takes the Sun about 220 million years to complete each revolution of the galaxy. It has been estimated – from theories of the Sun's formation and by dating the oldest Solar System material, such as lunar rocks and comets – that our Solar System formed about 4,500 million years ago, and the Sun itself must be about 5,000 million years old. This means that the Sun has orbited the galaxy about 23 times so far.

At first sight, the spiral shape of the galaxy tallies well with the fact that the galaxy is rotating: other rotating systems, such as hurricanes and water emptying down a plug hole, exhibit spiral shapes. In these cases, the spirals result from material near to the centre orbiting faster than material further out. However, things are not so simple with our galaxy: imagine a new-born Milky Way Galaxy with its four arms pointing straight outwards, with material orbiting faster the closer it lies to the centre. The arms would indeed wind up to give a spiral shape - but they would quickly become tightly wound. The spiral shape of the galaxy would have disappeared after only a few rotations, and yet even the Sun, not the oldest star in the system by far, has already been round about 20 times. The stability of the spiral shape (but not its origin) could be explained if the galaxy rotates more like a spinning rigid disc: in that case, material at the ends of the spiral arms would have to travel much faster than material near the centre to keep up. To see whether the galaxy rotates in this manner, astronomers have constructed a rotation curve: a graph of the orbital speed of galactic material versus distance from the galaxy's centre.

To work out the rotation curve of the galaxy, astronomers use a phenomenon known as the Doppler Effect, which is utilised, for example, in roadside speed cameras. If a source of radio waves with fixed wavelength is moving towards a detector, the detected wavelength will be shorter than if the source is stationary. Radio waves emitted by a source that is moving away will appear to have a longer wavelength. In the case of the speed camera, a car can be a 'source' of radio waves by virtue of the fact that radio waves reflect off the car's surface. In the case of the galaxy, it is the interstellar gas that is the source of the radio waves, with the very precise wavelength of 21.1cm. The rotation curve is constructed from measurements of the speed of galactic material at different distances.

The observed rotation curve does not fit the rigid disc scenario: orbital speed does not increase with distance; neither does it support the idea that the material nearer to the centre of the galaxy travels faster than material further out. In fact, the orbital speed is approximately the same for material at all distances of the galaxy. This result carries with it two conundrums. First, it explains neither the origin nor stability of the galaxy's spiral shape. Second, it suggests that the material of the galaxy is distributed fairly evenly throughout: if the material were more concentrated at the centre, rotational speed would decrease with distance out from the centre. (This is what happens in the Solar System, where the Sun, at the centre, holds more than 99% of the mass.) The distribution of all the known matter in the galaxy – stars, planets and the interstellar medium – is at odds with this interpretation. The inescapable conclusion, which makes sense of the rotation curve, is that there is a great deal of undetected material in the outer reaches of the galaxy. Astronomers and cosmologists call this hypothetical material 'dark matter'.

Galactic structure Astronomers picture the Milky Way Galaxy as consisting of three distinct components: the disc, the bulge and the halo. The disc is the spiral component of the galaxy. It is about 100,000 light years in diameter and about 2,000 light years thick. It is in the disc that newer, Population I stars are found. The disc is also home to most of the galaxy's interstellar gas and dust. The bulge is a near-spherical collection of mainly old, Population II stars about 32,000 light years in diameter. At its heart is the galactic nucleus, densely-packed with stars and gas. At the centre of the nucleus, and at the very centre of the entire galaxy, lies a black hole with a mass equal to two million Suns. This supermassive black hole is the final resting place of many stars and huge amounts of other material.

The third component of the galaxy is the halo. The globular clusters, whose distribution gave away the location of the galactic centre to Harlow Shapley in the 1920s, are distributed fairly uniformly throughout the halo. Most of the galaxy's older Population II stars are found within the halo; many of them in the globular clusters. With an observed diameter of at least 130,000 light years and a spherical shape, the halo totally encompasses the galactic disc. But the halo is thought to extend much further still, out to perhaps four or five times as far, and it is here that the galaxy's missing mass – dark matter – is thought to reside. The outer part of the halo is sometimes called the galactic corona; it is believed to contain perhaps four times as much mass as the disc, bulge and the rest of the halo combined! However, because it has not yet been detected, the nature of the unobserved matter in the galaxy's outer halo remains a mystery.

The dark matter in the outer halo of the Milky Way Galaxy cannot be made of stars, gas or dust, for these would certainly emit some kind of radiation – which, given that there is so much matter present, would have been detected. Several theories have been advanced to explain what dark matter might be. Some suggest that the halo is littered with small black holes; others that countless, very dim, brown dwarfs may populate it; others still that high-speed, long-dead neutron stars that escaped the galactic bulge long ago solve the mystery. The proposed black holes, brown dwarfs and neutron stars are collectively named massive astrophysical compact halo objects (MACHOs). Other theories suggest that the dark matter may be comprised of tiny particles that do not interact with normal matter other than through gravitational forces; such hypothetical particles are named 'weakly interacting massive particles' (WIMPs). Another possible contribution to dark matter may come from known, but hard-to-detect particles called neutrinos; more about those in Chapter Five. Whatever is the truth about dark matter, the efforts of many of the world's astronomers are focussed on finding out. One such effort, a search for MACHOs in the outer halo, is based on a concept called gravitational lensing, in which massive, compact, dark objects affect the light coming from more distant objects, distorting how we see the distant objects. I shall investigate this strange phenomenon in greater detail in Chapter Three.

The problem of dark matter is one of the most beguiling in modern cosmology. For far from being solely a solution to the problem thrown up by the rotation curve of the Milky Way Galaxy, the existence of dark matter is necessary if we are to understand the Universe as a whole. And so it is time to extend our census beyond our own galaxy, to the visible horizon of the Universe and beyond.

Beyond our galaxy I began this chapter by noting that the Universe appears isotropic from our viewpoint on Earth, and by suggesting that isotropy suggests homogeneity: on a grand scale, the Universe is the same throughout. That isotropy was spoiled by a milky-white band, the Milky Way, which turns out to be a 'local' feature: we are part of it. What would the sky look like if you could remove any local contribution to it, anything that is part of the Milky Way? For many astronomers working before the 1920s, the Milky Way Galaxy held all the matter in the Universe. If that were the case, then removing any intra-galactic contribution to what we can see would leave a featureless, black sky: completely isotropic but not very interesting. In fact, if you could do this, then to the naked eye at least the sky would indeed be completely black ... apart from three fuzzy patches of light.

These patches of light are separate galaxies outside the Milky Way Galaxy. Two of them are the Large and Small Magellanic Clouds (LMC and SMC), are 'companions' of the Milky Way Galaxy. They are visible from Earth only in the southern hemisphere. The Magellanic Clouds are small compared to the Milky Way Galaxy (the LMC is probably about 30,000 light years in diameter) and they are in orbit around it. The LMC contains about fifteen billion stars, the SMC about five billion. The third extragalactic patch of light is the Andromeda Galaxy, which is by far the most distant object visible to the naked eye. It lies about 2.2 million light years away and is a completely separate galaxy slightly larger than our own.

So much for the naked eye, with all its limitations. With the aid of technology – telescopes and photography – astronomers have shown that there are many, many more galaxies. In fact, the total number is estimated at around 100 billion. All the galaxies whose details can be seen – those closer to us – have stars, interstellar dust and gas, and supernovae that we can clearly observe. The strong suggestion is that other galaxies, though unthinkably far away, are made of the same matter, behaving in the same ways, as our own galaxy. This conclusion is supported by studies of other forms of electromagnetic radiation received from distant galaxies. Interstellar gas in other galaxies, for example, emit 21cm radio waves, just as it does in our own galaxy. Infrared studies of galaxies show up dust that has been heated by hot, young stars, so the amount of energy radiated as infrared is a clue to the rate of star formation inside a galaxy. Inside so-called 'starburst' galaxies, star formation is taking place at such a rate that the amount of infrared emitted is more than a hundred times the amount of visible light. X-ray studies of galaxies can bring to the attention of astronomers extremely energetic events. Some galaxies, called active galaxies, are dominated by an extremely bright nucleus that emits X-rays. Active galactic nuclei less than a light year across are thought to produce energy at the same rate as the entire Milky Way Galaxy. The source of an active galaxy's power is thought to be the hot accretion disc of a central supermassive black hole.

An important advance in our understanding of active galaxies came in 2000, when astronomers working with NASA's Chandra X-ray telescope announced that they had 'resolved the X-ray background'. For many years, X-ray telescopes had detected a fairly uniform, 'all-sky' emission of X-rays: the X-ray background. The Chandra telescope was the first to find any detail in the X-ray background, which it 'resolved' into a large number of individual sources. Each separate source turns out to be an energetic, active galaxy and is extremely far away; active galaxies are much more common very far away (long ago) than they are nearer to us (more recently). This suggests that many, perhaps all, galaxies were active when they were young, and quietened down with age, once their central black hole had consumed all of its available 'food'. There are probably supermassive black holes at the centres of most, or perhaps even all galaxies, but the black hole produces X-rays only when there is sufficient matter to 'fall into it' and create a hot accretion disc. The black hole at the centre of our own galaxy is starved of matter, and produces relatively little X-ray emission, for example.

Through powerful optical telescopes, astronomers can discover the shapes of galaxies. Many of the galaxies, including the Andromeda Galaxy, are spirals just like our own. Spiral galaxies that we see edge-on appear as we see our own galaxy: they are crossed by dark lines due to the extinction caused by interstellar dust. Furthermore, by studying the 21cm hydrogen line emission of other spiral galaxies, astronomers can construct rotation curves, just as they have done for our own galaxy. They can also work out the distribution of all the visible (luminous) matter in these galaxies; when they do, they discover that all spirals suffer from the same missing mass problem that ours does. Only about 25 percent of all known galaxies are spirals; about 70 percent are ellipticals: egg-shaped (ellipsoidal) or spherical conglomerates of stars moving around a central point in randomly-oriented orbits, not concentrated in a flat disc as with the spirals. The remaining galaxies – a few percent – are irregular in shape, not disc-shaped spirals or egg-shaped ellipticals. The Magellanic Clouds are examples of irregular galaxies.

The more distant a galaxy, the longer it has taken for its light and other electromagnetic radiation to reach us. When astronomers observe the most distant galaxies, they are seeing light and X-rays that left them nearly 13 thousand million years ago, when they were much younger. By comparing galaxies at different distances, then, astronomers can build up the life story of a typical galaxy. It turns out that young galaxies have different properties from older ones. By looking further back in time, astronomers have discovered that there were more dwarf galaxies when the Universe was much younger, and careful studies suggest that the shapes of galaxies we see today are the result of collisions and mergers between two or more galaxies. Starburst galaxies, in which star formation takes place at an exaggerated rate, are the result of galaxy mergers. Collisions and mergers between galaxies are not confined to the distant past: many examples of collisions have been identified closer to home. And in 1994, it was discovered that our own galaxy is undergoing a 'soft merger' with a dwarf galaxy. Astronomers have also found that the Andromeda Galaxy is heading towards us at a speed of about 300 kilometres per second. This means the two galaxies should collide in about 5,000 million years. But the Andromeda Galaxy is rare: only 120 or so galaxies are heading towards us. All of the other approximately 250,000 galaxies whose speeds have been measured are speeding away from us.

Shifting galaxies The fact that all the galaxies (except a few nearby ones) are hurtling away from us at high speed is clearly very important in our understanding of the Universe. To find out whether a galaxy is moving towards us or away from us, and at what speed, astronomers make use of the Doppler Effect. Rather than look solely at the 21cm hydrogen line, as they do when working out our galaxy's rotation curve, astronomers use any characteristic lines in the spectrum of radiation received from a distant galaxy, including visible light. When an object is moving away, the wavelengths of its light radiation become longer: they are shifted towards the red end of the spectrum. This has led to the term 'redshift'. A source that is moving towards us is said to be 'blueshifted'. The faster an object moves away from or towards us, the more its radiation is redshifted or blueshifted. Between 1912 and 1924, American astronomer Vesto Slipher (1875 - 1969) examined the spectra of 41 'spiral nebulae' (at the time, astronomers were unaware that these were separate, spiral galaxies) and found that 36 of them were redshifted, and so were receding at high speed. In 1924, American astronomer Edwin Hubble (1889 - 1953) discovered the truth about the spiral nebulae: that they are very distant, separate galaxies. In 1929, he made another startling discovery. He noticed that the fainter a galaxy appears, the more it is redshifted. In other words, the further away a galaxy is, the faster it is speeding away. This puzzling observation needed verification; the galaxies' redshifts were known, but what about their distances?

At the time, the only way astronomers could determine the distance to a galaxy was to measure the brightness (magnitude) of a standard variable star inside it, just as Harlow Shapley had done when measuring the distances to globular clusters in our own galaxy. However, this method was only accurate for galaxies that are relatively nearby: in more distant galaxies, the relatively dim variable stars did not stand out clearly enough to allow proper measurement of their magnitudes. Hubble extended the distance scale out to more distant galaxies using as his standard light source the magnitude of the brightest star in each galaxy he observed. The more galaxies he studied, the more evidence he gathered to support the idea that distant galaxies are speeding away faster than those that are closer. And to this day, for all except a very small number of galaxies, if you divide the speed of any galaxy by its distance from us, the answer is always the same: a number called the Hubble Constant. The proportionality between the distance of a galaxy and its speed is known as Hubble's Law. So well-tested and firmly established is Hubble's Law that astronomers actually use it to estimate the distances to galaxies that are too great to measure in any other way. In other words, the only way to determine the distances to the very furthest galaxies is by measuring their redshifts.

The determination of the exact value of Hubble's Constant is a major goal of astronomers and cosmologists, as we shall see in Chapter Six. However, with an approximate value, astronomers are able to estimate the distances to very distant galaxies, and use other, more accurate methods to measure the distances to the ones closer to us. This makes it possible to produce a three-dimensional map of the distribution of galaxies through the entire visible Universe. Astronomers have long known that galaxies exist in clusters, associated with one another by their mutual gravitational influence. Our own galaxy is part of a cluster called the Local Group, which has about 30 other members and has a diameter of about 10 million light years – a hundred times the diameter of our galaxy. But larger scale mapping brings out structures even larger than clusters: superclusters. Our Local Group of galaxies is part of the Virgo Supercluster, which has a diameter of about 100 million light years. And in 2014, astronomers discovered that even the Virgo Supercluster is part of an even bigger supercluster, which they named Laniakea.

At even larger scales, there is structure in the way the superclusters are arranged in space: they all look as if they formed along the boundaries of bubbles enclosing huge voids. By looking at the distribution of superclusters in space, cosmologists have confirmed that the Universe as a whole has a missing mass problem similar to the one in our galaxy. Dark matter, it seems, is to be found in the Universe at large, not just in individual galaxies. Inside the voids, no galaxies are found. On the very largest scale, the Universe looks something like a uniform foam of bubbles: homogeneous and something like a sponge after all. But if the Universe is so homogeneous, why are all the galaxies speeding away from us?

Expanding Universe At first, the fact that nearly all the galaxies in space are speeding away from us might seem to suggest that our own galaxy is somehow pushing them away. However, this is very unlikely: it would mean that our galaxy is very special indeed, whereas astronomers' observations tell us that all galaxies seem to be essentially the same, made of the same kinds of matter, for example. Furthermore, the strength of the hypothetical repulsive force responsible would have to increase with distance – and account for the fact that a handful of galaxies are moving towards us, and not away from us. One possible alternative explanation is that at some time long ago, there was an explosion in space, which sent all the galaxies hurtling away. However, that would require that our galaxy was exactly at the centre of the explosion: if the explosion occurred elsewhere in the Universe, then we would not see galaxies in every direction moving away. And still the mystery would remain of why the more distant galaxies are receding more quickly than those nearer to us.

The only satisfactory explanation for the facts of Hubble's Law is that the Universe – space itself – is expanding. The best analogy to illustrate the expansion of space is the inflation of a rubber balloon. Imagine white dots (galaxies) painted randomly on the rubber of the balloon (space). As the balloon inflates, the rubber stretches in all directions, increasing the distance between the dots. The further apart any pair of dots, the faster they move apart, because of the greater amount of stretching rubber between them. One important point to note here is that from the point of view of any dot, it would appear that all the other dots are moving away. Scaled up to Universe size, this analogy shows how Hubble's Law is not peculiar to our viewpoint in space, but would hold true for any galaxy. In an expanding Universe, the observed redshift of galaxies is partly caused by the stretching of electromagnetic waves as space through which they travel expands, an effect called cosmological redshift.

The fact that space is expanding implies that long ago, the entire Universe was much, much smaller. And that means the Universe must have been much hotter long ago, since the same amount of energy was then packed into a much smaller volume. In fact, using modern theories of matter, physicists can predict what the temperature and density of the Universe must have been and how matter would have been behaving at that time.

Detailed analysis of the conditions in this smaller, hotter, and much younger Universe is left to Chapter Six, but it is worth noting here that astronomers have detected one clear 'fingerprint' of that long-lost Universe: radiation called the cosmic microwave background (CMB). Like any matter, the matter in the hotter, younger Universe would have been emitting 'thermal radiation': a range of wavelengths of electromagnetic radiation, the exact spectrum of which is dependant on temperature. Above a certain temperature, atoms are not stable, and the matter of the hot, younger Universe would have existed as a seething soup of charged particles, with the thermal radiation being constantly absorbed and re-radiated. As the Universe expanded, it cooled, and at a temperature of around 3000 Kelvin (about 2700°C), atoms would have become viable entities at last. With stable atoms present, the radiation was not now constantly absorbed and re-radiated, but could travel freely across space. Importantly, the spectrum of the radiation would have been characteristic of the Universe's average temperature at the time, 3000 Kelvin. And some of it still survives today, as the CMB radiation: an imprint of the 'epoch of recombination', frozen in time. It does indeed have a spectrum characteristic of thermal radiation at a particular temperature. However, the 'temperature' of the cosmic background radiation is not 3000 Kelvin: instead, it is an extremely cold 2.725 Kelvin, less than 3 degrees above the lowest possible temperature (absolute zero; 0 Kelvin, -273°C). The explanation for this lies in the cosmological redshift: the radiation has been 'stretched out' as space has expanded. At the time of recombination, the Universe's average temperature really was 3000 Kelvin. Another way to look at this is that the Universe has indeed cooled as it has expanded, and the background radiation is indicative of its much cooler average temperature today.

And so, as we come to the end of our review of the census of the Universe, it is possible to summarise it thus:

We live on a planet, orbiting a star, which is in a galaxy, which is part of a cluster, which is part of a supercluster, which lies at the edge of some kind of bubble about 1000 million light years in diameter, which surrounds an enormous void in a foamy Universe which is continuously growing.

However satisfyingly concise that summary of our situation may seem, it generates as many questions as it answers. What actually is space? How can it expand? What is time? What is matter made of? When and how did the Universe come into existence? What is the source of the large scale structure of the Universe? In what sense is the Universe real? In the chapters that follow, I shall investigate these questions, beginning with scientists' attempts to unravel the mysteries of space and time.

2: SPACE AND TIME

Two of the most pertinent questions about the reality of our Universe are: "What is space?" and "What is time?" These questions sound simple enough, if perhaps a little philosophical. Some equally straightforward answers might be: space is the physical arena in which matter exists and in which things happen; time is what makes one moment different from the next – or, as physicist John Wheeler suggested, "what keeps everything from happening all at once". As we are about to discover, neither space nor time can be defined so simply. Modern physicists consider space and time to be intimately woven together, forming a four-dimensional landscape called spacetime. To see how they arrived at this, and just what it means, we need to examine how theories of space and time have developed.

Space – 'nothing' really matters It is tempting to think of space as simply emptiness: matter can take up space, but elsewhere, space just is. According to the *atomic theory*, the particles that make up matter – atoms and molecules – are tiny compared with the amount of space available. They exist in a void. Furthermore, within each atom there is space between the nucleus and its attendant electrons. So even when space is occupied by solid matter, only a tiny fraction of it is actually 'taken up'. The rest is not made of anything; it is nothing. How, then, can we explain the well established observation that the Universe is expanding? The stars and galaxies are not themselves growing in size. Nor are the galaxies flying apart into surrounding space: it is the space between the galaxies is expanding. How can 'nothing' expand?

Debates about the nature of space began in ancient times. Some Ancient Greek philosophers – notably Democritus (c. 460 - 370 BCE) – proposed the first well developed atomic theory, which included the notion of void. Others considered the idea of truly empty space to be unacceptable. Parmenides (c. 510 - 440 BCE) argued that if empty space really is nothing, then there is nothing between the particles of matter, and they cannot be any separation between them. He suggested instead that all of space was suffused with a material which he called plenum, meaning 'full'. A century or so later, Aristotle made his famous declaration that "nature abhors a vacuum". This is another way of saying that the Universe is full – he agreed with Parmenides. Aristotle's claim was based on logical deduction, and also on his observations of suction – gases and liquids flowing to fill otherwise empty space. He, too, suggested the existence of a material that filled space, which he called aether.

The philosophical wrangling about empty space became significant again during the Renaissance in Europe. Influential mathematician and philosopher René Descartes (1596 - 1650), for example, agreed with Aristotle. He suggested that even what we call empty space must be some kind of substance. Just as an urn is said to be empty when it contains only air, empty space must still contain something. Substance can be identified by the fact that it has 'extension' – in other words, you can measure its length, breadth and width. You can also measure the extension of space, so space must be a substance: "Since it has extension, there must be substance in it." Descartes' Aristotelian notion of the non-existence of the perfect vacuum was challenged by several of his contemporaries. Most notable among them was Italian physicist and mathematician Evangalista Torricelli (1608 - 1647), famous for establishing the notion of atmospheric pressure.

Torricelli suggested filling with liquid a long tube closed at one end, and standing it vertically with its open end immersed in a bowl of the same liquid. According to Aristotle's view, the tube would remain filled with the liquid, since otherwise a vacuum would be left at the top. When Torricelli's colleague Vincenzo

Viviani (1622 - 1703) carried out the experiment, he found that Aristotle was right only up to a point. All liquids have a maximum height to which such a column can be sustained. Above that height, if the tube is long enough, a void appears. Torricelli's (correct) interpretation was that it was the pressure of the air pushing down on the surface of the liquid in the bowl that supports the liquid. Atmospheric pressure can support only a certain weight of liquid, so the height of the column depends upon the density of the liquid involved: the more dense the liquid, the less high the column could be. Toricelli had invented a barometer – a device that could measure atmospheric pressure. Water in this kind of barometer rises to about ten metres (35 feet), while mercury – the most dense liquid commonly available – rises to about 76 centimetres (30 inches). Both heights depend on atmospheric pressure, which changes slightly over time.

Inspired by Torricelli, German politician, inventor and experimenter Otto von Guericke (1602 - 1686) made (partial) vacuums by sucking air out of various containers with a hand-held pump. After many improvements to his pump, he carried out a host of experiments with vacuums, including proving that sound can't pass through them – a fact that seemed to support the idea that space is empty, since sound requires a physical medium through which to pass.

In 1654, von Guericke staged a spectacular public demonstration in his home town of Magdeburg. He held two copper hemispheres together to form a sphere 60 centimetres (24 inches) in diameter. He made the sphere airtight with leather and wax and evacuated the air from inside with his pump. With no air inside the sphere, there was no air pressure pressing outwards to balance the atmospheric pressure on the outside of the sphere – and the atmospheric pressure pushed the two hemispheres tightly together. So tightly were the hemispheres pushed together that two rows of horses could not pull them apart. But when von Guericke opened the valve to let air back in, even a small child could easily pull the hemispheres apart. The following year in England, Irish-born scientist Robert Boyle (1627 - 1691) read about von Guericke's research, and set about designing an improved vacuum pump. His assistant, English scientist Robert Hooke (1635 - 1703), helped with the design and construction, and together they built a hand-cranked, piston-driven pump that was much more powerful.

In inventing the barometer, Torricelli had apparently succeeded in making a perfect vacuum: the void above the column of liquid. Atomic theory, which was then about to become part of mainstream science, actually puts a damper on Torricelli's claim of creating a true void. One of the many strengths of the atomic theory is that it explains evaporation, as the escape of individual atoms or molecules from the body of a liquid. Under that interpretation, the space above the level of mercury inside a barometer is actually filled with a rarefied vapour consisting of mercury atoms, and not a perfect vacuum. Despite refuting Torricelli's claim, of course, the atomic theory actually supports the concept of the perfect vacuum, since – as I've already pointed out – atoms are extremely tiny, self-contained objects with nothing but empty space between them. So perhaps Torricelli's "vacuum" was empty space after all – just peppered throughout with tiny particles.

English physicist Isaac Newton (1642 - 1727) was a great believer in the atomic theory: he considered everything, even rays of light, to be made of particles. Despite his atomic standpoint, Newton did not like the idea that atoms existed in otherwise empty space. He toyed with the idea of an all-pervading substance which he called the aether, like the aether of Aristotle and the plenum of Parmenides. Part of the reason for Newton's belief in the aether was the apparent nonsense of the idea that objects separated by empty space could have any influence on each other. How could the forces of gravity between the Sun and the planets be transmitted across empty space, for example? Despite this problem, Newton supposed that even the aether would have to be composed of particles, so he was left with the problem of what could fill the gaps in between the aether particles. The only answer seemed to be "empty space".

The 18th century saw growing acceptance of the atomic theory, and the idea of empty space gradually became more acceptable. Around the beginning of the 19th century, though, an idea arose that was to spell trouble for the void: convincing evidence was mounting that light is a wave motion. Waves are undulations within substances: water waves are undulations in water, sound waves are undulations in air. And yet, unlike sound, light was found to travel unimpeded through the scientists' laboratory vacuums. So the 19th century scientists proposed that some kind of aether must exist throughout empty space, because light waves must be undulations in it. Cautiously, perhaps, they restricted the existence of the aether to a substance through which light could pass: the luminiferous aether. Whatever it was made of, the luminiferous aether would have to be transparent, massless and continuous. The quest to determine the nature of the hypothetical aether was made more daunting in the 1870s, after a monumental discovery by Scottish physicist and mathematician James Clerk Maxwell (1831 - 1879). Maxwell had formulated equations that describe the behaviour of electricity and magnetism; when he combined his equations, he found they described a wave travelling at exactly the speed of light, which had been measured not long before. The inescapable conclusion was that light (and other forms of electromagnetic radiation, which Maxwell's equations predicted) travels as waves of varying electromagnetic fields. Interestingly, they are transverse waves – which means the oscillations are at right angles to the direction of travel, like waves on a pond. In fact, transverse waves are observed at the surfaces of solids and liquids, not inside them; where they do travel inside a substance, that substance has to be a solid. Assuming we live within space, then, the luminiferous aether must be solid. The aether had to be not only transparent, massless and continuous, but rigid, too! Explaining how the stars and planets could move through such a medium clearly presented a serious problem.

Maxwell's theory of light as electromagnetic waves was based on the idea of electric and magnetic fields. The idea of a 'field' of force underpins much of today's understanding of matter, as we shall see in Chapter Four. Field theory originated with English physicist Michael Faraday (1791 - 1867), who carried out experiments to investigate the link between electricity and magnetism. Faraday's observations led him to postulate the existence of force fields around magnets and electrically charged objects, and he went on to give shape to the invisible electric and magnetic fields. Using iron filings, he was able to highlight the 'field lines' that carried the electric and magnetic forces. But were the fields arrangements of the particles of the aether? Or did the fields somehow stretch through empty space?

To solve the problem of the aether, 19th century physicists invested a great deal of time and effort into proving its existence. One of the most important, and best-conceived, experiments was carried out many times during the 1880s, by Polish-born physicist Albert Michelson (1852 - 1931). The experiment was designed to determine the speed of Earth through the aether. Moving through the aether towards a light source, the speed of light would be measured as greater than if you are moving away from the light source. Michelson's experiment would detect any difference in the speeds of two beams of light at right angles to each other. But, to the physicists' surprise, no difference was found. The speed of light was exactly the same whether you were speeding towards a light source or away from it. The suggestion seems to be that there is no aether. This is a problem for the electromagnetic field theory, since transverse light waves clearly reach us from distant stars, crossing empty space.

Although modern physicists' notions of space are more sophisticated than they were back then – as we shall see – its true nature remains a mystery to this day. One thing is certain: empty space is certainly not empty. The well-tested theories of quantum physics developed in the twentieth century reveal that it is seething with 'virtual particles' that pop fleetingly into and out of existence. I'll delve into the 'quantum vacuum' in Chapter Four. But the mysterious null result of Michelson's experiment was eventually solved by Einstein's Special Theory of Relativity, which also revolutionised the concept of time. Before explaining the view of space presented by Einstein's theories, then, let's take some time to investigate the meaning of time.

Time: a one-way street The concept of time, and in particular the measurement of its passing, have long been fascinations of the human mind – and of course practical considerations in every society. Calendars are based on annual cycles. The Sun and the stars shift their positions gradually day-by-day – a cycle that repeats each year. This phenomenon is, of course, due to the fact the Earth takes a year to complete each revolution of its orbit around the Sun. The Ancient Egyptians famously used the annual appearance of the bright star we now call Sirius above the horizon as a sign that the waters of the Nile delta were about to flood, spreading fertile soil over their agricultural land. This made it possible to track time year-by-year, but attempts to divide the year into smaller fragments were fraught with difficulty. This is because the year does not consist of an exact number of days – why should it? The year is, in fact, approximately 365-and-one-quarter days long. The best that early civilisations could do was to add or subtract days every so often to keep their calendars in line with the advancing years. The Julian calendar, adopted by order of the Roman emperor Julius in the year 46, introduced a leap year every four years, the extra day making up for the extra quarter of a day per year. This system was improved in 1582, by order of Pope Gregory XIII, by refining this concept of leap years. The Gregorian calendar is more accurate than the Julian one, but it still ultimately suffers from the same problem.

The one measure of time that could always be depended upon was the day, which is based on the rotation of planet Earth. Once each day, at 'local noon', the Sun reaches its highest point in the sky. The next noon occurs 24 hours – one 'solar day' – later. Based on this definition, one second is exactly one-eighty-six-thousandth of a day. Unfortunately, however, the speed Earth moves along its orbit changes during the year, and this affects the length of the solar day. This is why 'mean time' was invented: a 'mean solar day' is the length of time from noon to noon averaged over the year. However, even this is not steadfast: one day is longer by 0.002 seconds nowadays than it was a century ago, for example. And so in 1967, the Thirteenth General Conference on Weights and Measures redefined the second as "the duration of 9.192,631,770 periods of the radiation corresponding to the transition between the two hyperfine levels of the ground state of the caesium-133 atom".

Scientists have developed the technologies to measure time extremely accurately, but what can they say about the nature of time itself? Time can sometimes feel like it is passing more slowly or more quickly, but it certainly seems to 'flow' in one direction only. We have memories of the past, but not of the future; we are apparently able to influence events in the future, but not those of the past; and we are truly conscious only of the moment we call 'now'. Does the past exist in any real sense, or is it only in our memories? Does the future already exist, or is it continuously being created, like the crest of a wave that carries us along? These are simple yet deeply challenging questions.

In 1927, English physicist Arthur Eddington (1882 - 1944) named the direction of time's flow "the arrow of time" – a metaphor that has become part of scientists' vocabulary. The arrow of time can be defined by causality: an event that happens at a particular time can only have been caused by events that happened prior to it – at an earlier time. Simple as this may sound, causality is not as straightforward as it seems. Almost all individual physical processes are symmetrical in time: equally possible in reverse as forwards. In other words, the laws of physics, when expressed mathematically, work just as well when you substitute '-t' in place of 't'. For example, if you could watch a video of individual atoms colliding, you would not be able to discern whether the video was being played forwards or in reverse. However, there are clearly everyday occurrences that would look very odd if you watched them played on a video in reverse: a pool player hitting a cue ball into a pack of balls is a good example. Strange as it may seem, however, there is no law of physics that says this scenario is impossible: the balls, in what we think of as their final positions, could spontaneously absorb energy from their surroundings, making them move in just the right directions to converge in a pack, transferring their energy to the cue ball, which would impact the pool player's cue with just the right amount of energy to make his or her arm jolt backwards. It sounds like nonsense, but nature does not outlaw it.

So why do pool balls never spontaneously converge? What is it that pushes the arrow of time in one direction and not the other? Since the end of the 19th century, physicists' answer to this question has relied on scientific insight provided by an area of physics called thermodynamics. There are three well-tested laws of thermodynamics, each of which deals with the transfer of energy. The first law of thermodynamics simply states that energy is never lost nor is it ever created. The energy given to the cue ball in the example above is transferred, by impact, to the balls in the pack, which then move off at speed. As a result of the collision, the cue ball slows down to a stop, but its energy is not destroyed. Most of its energy is transferred to the other balls, and the rest has been converted into sound and heat energy. The energy of the other balls is ultimately converted into heat and sound, too. The point is that all the energy can be accounted for. And, since on a microscopic level, the processes involved in this scenario are reversible, it is worth stressing again that it is possible – from the point of view of energy conservation – that the whole series of events could happen in reverse. However, the second law of thermodynamics explains why this time-reversed scenario has never been observed. The second law deals with a concept called 'entropy', which is a measure of the disorder of a system. Entropy, it states, is always rising: it increases with time. (The third law also involves entropy, but does not include any reference to time.)

To understand what entropy has to do with time, consider the pool table with the coloured balls in their starting position and the cue ball hurtling towards them. The cue ball possesses some kinetic energy, by virtue of its motion towards the coloured balls. The state of the pool table a few seconds later is clearly more disordered: the balls are now spread randomly around the table. But there is something more poignant happening at the microscopic level, which illustrates the importance of entropy in any situation. To understand what, we must remember that the balls are composed of molecules, joined together in a rigid structure. The molecules of each ball are all vibrating – not violently enough to break free from each other, or else the ball would become a liquid or a gas – but they are all vibrating, randomly. The temperature of a ball, and indeed any substance, is directly related to the average energy of its component particles. Increasing the average vibrational speed of the ball's molecules is the same as increasing the ball's temperature. When the cue ball hits the first ball, some of the kinetic energy is taken by the molecules of the ball, resulting in an increase in their vibration. This extra kinetic energy is shared out randomly between all of the molecules, and the ball's temperature rises very slightly. This happens during each collision of the balls, and during each collision of a ball with the pool table's cushions, and in each case, a little heat is generated. The sound waves produced by the collisions are disturbances of the air around the balls, and the energy of these disturbances, too, is eventually incorporated into the air and the walls of the pool room, again causing a slight increase in temperature which is quickly and evenly distributed.

The initial kinetic energy of the cue ball is ordered, and is therefore 'available': it can be used to move the other balls. The vibrational energy of molecules within the ball is random, disordered, and is quickly shared out so that it can no longer be used. Entropy, then, is really a measure of the 'unavailability' of a system's energy. When substances are in thermal equilibrium – at the same temperature as each other – their thermal energy cannot be used to do anything. And matter always tends to reach thermal equilibrium as time moves on. An ice cube dropped into warm water will always melt to form liquid water, simply because the vibrational (thermal) energy of the water molecules is shared between the ice and the liquid water – ultimately leaving slightly cooler liquid water. Temperature is related to the average energy of the molecules of a substance, but some molecules have a little energy, others much more. There is a range of energies. It remains possible for millions of the 'coldest' (lowest energy) water molecules to be in the same region, and to form ice. But it is so extremely unlikely that it is never actually observed.

It is worth noting that there are, of course, many processes in which entropy appears to be reduced: for example, a refrigerator cools its contents below room temperature. In this case, however, there is an input of energy to the system: electrical power. The refrigeration process actually produces a net heating effect – you can feel this at the back of a refrigerator. This heat is released to the room and shared randomly between the air molecules. If you leave a refrigerator door open and turn on the refrigerator, the room will grow slightly warmer, not cooler. The entropy of the whole system increases. There are countless other examples in which entropy decreases locally, but at the expense of an equal or greater increase elsewhere. The existence of life, for example, seems extremely unlikely: the probability that all the molecules of your body are in the right arrangement is extremely small, so living things are islands of consistently very low entropy. However, this does not conflict with the second law of thermodynamics, since making and maintaining living things requires an input of energy, which ultimately ends up as evenly-distributed heat.

It is a statistical near-impossibility for pool balls to move spontaneously or ice cubes to form in warm water. So, it is a causal law of nature (if only a statistical one) that entropy increases with time. Every energy transfer produces a little heat, and heat is eventually shared out in time, achieving thermal equilibrium. This provides scientists with a comforting, well-defined direction for the arrow of time. It may also explain the 'psychological arrow' of time: the conscious awareness we have of time's flow. Although there is still much for neuroscientists to discover about the human brain, it is clear that brains are powered by chemical reactions

– physical processes that ultimately involve an increase in entropy.

Another manifestation of thermodynamic entropy is the 'cosmological arrow' of time. We saw in Chapter One that the Universe is expanding. At the epoch of recombination – about 300,000 years after the Big Bang, when atomic nuclei and electrons combined to make atoms – the Universe was smaller and more ordered. Although the Universe's average temperature was much higher then, the entropy was low, because there was a great availability of energy. As the Universe has expanded, much of the energy has been converted, ultimately, into heat. As stars shine, they are converting 'useful' gravitational energy into radiation. This radiation interacts with matter, and as time moves on, any 'hotspots' or energy-producing regions, will eventually reach thermal equilibrium with the rest of the Universe. This concept – the disordered, lukewarm future of the Universe – has been called 'heat death'.

Having stated that the laws of physics work equally well in both directions, I need to point out that there are exceptions. Until recently, all of the known laws describing individual interactions between subatomic particles - including collisions, radioactive decay, emission of radiation – were indeed symmetrical in time. But in 1998, an experiment at CERN Laboratories in Geneva did find direct evidence of a process that is not symmetrical in time. The experiment involved producing streams of subatomic particles called kaons and their partner particles, anti-kaons; kaons can change into anti-kaons and vice versa. We shall encounter particles and antiparticles in Chapter Five, but for now it is enough to say that the results of the experiment showed that kaons change into anti-kaons more often than the other way round. The point is that the two reactions are the reverse of each other in time, so the experiment highlighted for the first time an example of time-asymmetry. Since then, other examples if time asymmetry have been found in the world of subatomic physics. Time really does run in one 'direction' and not the other.

Space and time: moving ideas The concept of motion relies firmly on concepts of both time and space. It is no surprise, then, that it is the study of motion which can tell us most about the nature of time. In ancient Greece, philosophers were troubled by motion. Zeno of Elea (5th century BCE), for example, believed that motion is impossible, and must be an illusion. He put forward his arguments in the form of paradoxes. One of these involved an arrow shot from a bow. At any particular moment, the arrow is in one particular place; at all times, therefore, it is stationary, not moving. A similar conclusion is reached by considering a line as an infinite collection of points: no object can move through an infinite number of points in a finite time. These are powerful arguments – modern solutions to Zeno's paradoxes are mathematical and very complicated. But they are not very useful if you are hoping to discover the true nature of space, time and motion. Aristotle criticised Zeno, saying that common sense tells us that motion must be possible, because it is observed. Common sense can often prevail, but it can be a pitfall, too.

Aristotle's world view suffered from many common sense misconceptions. One of them had great significance in the development of theories of space and time. Aristotle believed that Earth was fixed firmly at the centre of the Universe, with the stars and planets circling around it. This carried with it the idea of 'absolute rest' – any motion could (in principle at least) be measured relative to Earth, unmoving at the centre of the Universe. So compelling and metaphysically satisfying was this idea that it was widely held. It also appears in Psalm 104 of the Bible, for example: "He set the earth on its foundations; it can never be moved". Aristotle claimed that if the Earth were spinning and moving through space, we would be thrown about, or at least notice some consequences in the movements of objects. For example, he supposed that objects dropped from a great height would be seen to swerve to one side, rather than fall straight down. The Earth-centred, or geocentric, view of the Universe ruled for nearly two thousand years, and was promoted as fact by the Catholic Church, who adopted the views of Aristotle and, of course, the Bible.

In 1632, the Italian scientist and mathematician Galileo Galilei (1564 - 1642) dared to challenge the geocentric view publicly, in a book of dialogues entitled *The Dialogue on the Two Chief World Systems*. The book illustrated the folly of Aristotle's ideas. Through the dialogues, Galileo suggested that Earth is in orbit around the Sun, which would mean that it was neither at the centre of the Universe nor in a fixed position. Galileo had been convinced of the sun-centred (heliocentric) view of the Solar System since about 1610, after making careful observations using telescopes, and as a result of reading a book written by Polish astronomer Nicolaus Copernicus (1473 - 1543), called *De Revolutionibus Orbium Coelestium* ("On the Revolution of Celestial Orbs"). Copernicus had been the first Western scholar since ancient times to promote the heliocentric system, which he had devised while employed by the Church to improve the calendar system. Copernicus realised that the only way to improve the accuracy of the calendar was to assume that the planets – including Earth – revolve around the Sun. Copernicus' book, published when he was on his deathbed, did not attract much attention until Galileo published his dialogues promoting Copernicus' ideas. Not only did the book challenge the accepted world view, it parodied the dogmatic approach of the Catholic Church. Galileo was put under house arrest. His book was put on the Church's Index of Prohibited Books, where it remained until 1824. The Catholic Church finally made an official apology to Galileo in 1993.

In one of the most important parts of his dialogues, Galileo considered the motions of people and other objects aboard a ship. Among his examples were water dripping vertically from one bottle into another and an object thrown from one person to another. He rightly claimed that there would be no differences in these motions if the ship were stationary or moving at a constant velocity (unchanging speed and direction). The water still drips vertically and the object does not have to be thrown any harder if the ship is moving.

Galileo's point can be expressed in the following way: 'the laws of physics remain unchanged when moving at a constant velocity'. This has become known as the Galilean Principle of Relativity. It was important in showing how Earth could be hurtling around the Sun and spinning once a day without us noticing anything strange.

Fortunately, the Catholic Church's Index of Prohibited Books had little influence beyond Italy, and the new ideas began to spark interest in the heliocentric system throughout the rest of Europe. In England, Isaac Newton adopted Galileo's views, and formalised them in his laws of motion. He did not, however, give up the idea of absolute rest: all motions may only be measured relative to each other, but they are still relative to 'fixed' space. In one of his most important scientific works, *Philosophiæ Naturalis Principia Mathematica*, Newton wrote: "Absolute space, in its own nature, without relation to anything external, remains always similar and immovable." Newton was stating his firm belief that space is a fixed framework: aether or no aether, space is an arena in which matter can exist. In the same publication, Newton stated that: "Absolute, true, and mathematical time, in itself, and from its own nature, flows equably without relation to anything external". In Newton's view, then, although time is measured relative to the seasons, day and night, and ticking clocks, these are all reflections of some 'absolute time' ticking away the moments of the Universe. With these ideas of space and time firmly established, Newton formalised his mathematical laws of motion, which very accurately describe the way objects move under the influence of forces, and indeed when no net force is acting.

Einstein's motion Newton's ideas of absolute space and absolute time, and therefore the idea of absolute rest, were shown to be false by the work of Austrian-born physicist Albert Einstein (1879 - 1955). In his 1905 paper *On the Electrodynamics of Moving Bodies*, published in the German scientific journal *Annalen der Physik und Chemie*, Einstein presented his Special Theory of Relativity. The theory is based on two postulates. The first is that the laws of physics should be the same for observers in uniform motion – movement at a constant velocity. Anyone in uniform motion can consider himself or herself to be at rest, and all other motions relative. This says essentially the same as Galilean Relativity, but Einstein had to extend it because of his knowledge of the laws of electromagnetism, which were unknown to Galileo and Newton.

The second postulate of Special Relativity, then, states that the behaviour of electromagnetic radiation is correctly described by James Clerk Maxwell's theory of electromagnetism (discussed earlier). Since Maxwell's theory predicts that electromagnetic radiation is a wave with a clearly defined speed, the two postulates taken together demand that the speed of light (and other forms of electromagnetic radiation) should be measured as having the same value for all observers in uniform motion. So, travel towards a lamp at half the speed of light, and the lamp's light will approach you at the same speed as it would if you were stationary with respect to the lamp, or even moving away from it. This is why Michelson's experiment found no evidence of the aether. The two postulates of Special Relativity have a number of strange consequences that were to shake the foundations of the scientific world view of the time. One of the most disturbing consequences of Special Relativity was that the concept of simultaneity is relative: two events that happen at the same time from the point of view of one observer are not simultaneous from the point of view of a second observer moving relative to the first.

To understand why this is so, consider yourself standing still at the midpoint between two light sensors that are one light-second apart (300,000 kilometres). If you send out a light pulse in each direction, you assume that they will reach the sensors simultaneously. To make sure, the light sensors automatically send out a flash of light the moment they receive your light pulse. You see the two return flashes simultaneously, as you would expect. Now imagine the scene from the point of view – or 'frame of reference' – of someone moving relative to you at high speed. As far as they are concerned, they are at rest and you are moving. So, one light sensor is moving towards the point at which your light pulse originated, the other one is moving away from it. This means the two pulses travel different distances – and, since the speed of light is absolute, arrive at different times.

Measurements of time – in fact our whole everyday conception of time – is based on the notion of simultaneity. As Einstein pointed out in his 1905 paper, "If, for instance, I say 'The train arrives here at seven o'clock,' I mean something like this: 'The pointing of the small hand of my watch to seven and the arrival of the train are simultaneous events.'" In Newton's view of space and time, a specific moment is the same for the entire Universe. If you view time in this way, you could represent the history Universe as a series of fixed snapshots – like frames of a universal, three-dimensional video sequence. All simultaneous events anywhere in the Universe will appear on the same frame. Special Relativity destroys the notion of universal simultaneity, so that – to return to the analogy – observers moving relative to each other would have differing snapshots of the universal video. Events that appear in the same video frame for one observer will be in different frames for another, and may also be in different positions in the frame. This bizarre effect is only noticeable at a very high, or 'relativistic', speed – a significant fraction of the speed of light – which explains why Newton was unaware of it.

Einstein went on to show that the absolute nature of the speed of light also dictates that clocks carried by two observers in relative motion run at different rates. A 'stationary' observer will observe the clock of a 'moving' observer run more slowly. This does not mean that the clock becomes faulty when it moves, but that time itself literally runs at different rates in different frames of reference. All physical processes will appear to happen more slowly in the moving frame of reference relative to the stationary one – including biological ones, such as ageing. Put simply, moving clocks run slow – an effect called time dilation. Stranger still, since there is no such thing as absolute rest, each observer sees himself or herself as stationary and the other as moving. And so, each observer feels time passing at an unchanging rate. Within your own frame of reference, one second will always be one second, however slowly your time appears to travel as measured by an observer moving rapidly past.

The relativity of time is completely symmetrical – neither observer is 'wrong', there is just no 'right'. At first sight, this gives rise to a paradox, since two people moving apart at a relativistic speed will both see the other person ageing more slowly than themselves. This is explored in the well-known 'twins paradox'. In this thought experiment, one of a pair of twins sets off at relativistic speed from Earth, travels for several years to visit a distant star, and then turns back and returns to Earth, again at high speed. When the twins are reunited, they will have aged differently. This may be a thought experiment, but if this scenario was played out in the real Universe, the twins really would age at different rates. The resolution of the paradox is complex; the simplest explanation depends upon the fact that observers moving relative to each other measure not only times differently, but distances differently, too. Distances parallel to the relative motion, such as the distance from Earth to the star, will be measured as shorter by the 'moving' observer than the 'stationary' one. This is called length contraction, and is easy to deduce directly from the absolute nature of the speed of light. Measure a stationary rod as one metre long, and it will be shorter if you measure it when it is moving relative to you at high speed. So, the twin who travels to the star and back travels a smaller distance than is measured from Earth, taking less time to make the trip than the twin on Earth measures. The twins do age differently. Einstein illustrated length contraction with the example of a sphere moving past you at relativistic speed, which will appear flattened in the direction of motion. So, if someone was moving past you at relativistic speed, you would see their body flattened in the direction of motion. Again, since each observer sees himself or herself as stationary and the other as moving, this relativity of distance is symmetrical, and neither observer is wrong.

In his presentation of the theory of Special Relativity, Einstein showed that the speed of light is not only a universal constant, but is the Universe's speed limit: nothing can travel faster than electromagnetic radiation. He proved this in three different ways. One is a direct consequence of the length contraction: in the example of the flattened sphere above, the equations of Special Relativity show that the contraction would be infinite if the sphere were moving past you at the speed of light – clearly, it would be nonsense for the sphere to travel any faster. The second proof is an extension of the relativity of distance to the relativity of speed – a fairly trivial mathematical operation – which shows how two observers each travelling at, say, three-quarters (0.75) of the speed of light towards each other will measure their relative speed as less than the speed of light. Common sense would suggest that the figure should be 1.5 times the speed of light (0.75 + 0.75), but this would not allow for existence of electromagnetic waves.

The third proof that the speed of light is the Universe's speed limit involved equations that describe the motion of an electrically-charged object. Einstein constructed mathematical equations that calculate the energy required to accelerate such an object. According to Newton's Laws of Motion, the amount of energy needed to increase the speed by a certain amount depends only upon the mass of the object. So, to accelerate a heavy lorry from 50 to 55 kilometres per hour requires the same energy input as is needed to accelerate it from 55 to 60 kilometres per hour. Einstein's equations showed that Newton's Laws are incorrect at relativistic speeds. The amount of energy needed to increase the speed of an object by a certain amount is greater the higher the speed. It turns out that to accelerate an object to the speed of light would require an infinite amount of energy.

Later in 1905, Einstein published a separate paper, entitled 'Does the Inertia of a Body Depend Upon its Energy Content?'. The term 'inertia' is generally taken as having the same meaning as 'mass', since it is a measure of the resistance to a change in velocity. In this second document, Einstein further developed his relativistic energy equations, and he was able to show that the inertia of an object is reduced if it loses energy in some way, such as by emitting radiation. He concluded that "the mass of a body is a measure of its energy content" and he speculated that if this were true in practice as well as in theory, then objects that do lose energy, such as radioactive substances emitting gamma rays, may also lose mass. In 1907, Einstein published a review of relativity theory, in which he summarised these ideas in his most famous equation:

$$E = mc^2$$

The c in the mass-energy equation refers to the speed of light, the m refers to the mass, and the E to the total energy.

Einstein's equation can easily be misunderstood. It seems to suggest that an object's mass increases as its relative speed increases: that mass is a relative quantity. However, it is the total energy of an object that is the relative quantity; the mass remains the same. It is true, though, that when an object is at rest relative to an observer, its mass, sometimes called the 'rest mass', is equivalent to an amount of energy. The c in the mass-energy equation, the speed of light, is a very large number. So an object with a mass of just 1 gram (0.001 kilograms) is equivalent to an energy of 90 trillion joules – this is enough energy to keep a 10 watt bulb alight for nearly 300,000 years. Einstein's mass-energy equation is used routinely by particle physicists. For example, when pairs of particles annihilate each other, producing radiation, the energy of that radiation is equal to the total mass-energy of the annihilated particles.

Space time geometry Special relativity is concerned with 'events'; an event is anything that can be described by a point in space and a moment in time. Three co-ordinates are needed to define the position of a point in space; that is why we are said to inhabit three-dimensional space. The easiest way to visualise what Special Relativity says about simultaneity is to consider including time as an extra dimension, on an equal footing to the dimensions of space. German mathematician

Hermann Minkowski (1864 - 1909) suggested doing this in 1908. With three space dimensions and one time dimension, Minkowski successfully formulated Einstein's theory in terms of a four-dimensional co-ordinate system called spacetime. This is the perfect mathematical arena in which to describe the effects predicted and explained in the Special Relativity theory.

Every observer moves at exactly the same speed through spacetime: the speed of light. In everyone's own frame of reference, all of that speed is concentrated along the time dimension: we are all moving along our own time dimension at a constant speed. While you trundle along your time dimension at the speed of light, someone in relative motion to you will be travelling along the dimensions of space (at least the dimensions of space in your frame of reference). Since their overall speed through spacetime must still be exactly the speed of light, not all that motion can be along the time dimension (in your frame of reference). In other words, they are travelling along the (your) time dimension at less than the speed of light – time passes more slowly for them, from your point of view. Of course, as pointed out above, the same is true for you from their frame of reference.

Considering time as a dimension through which we move, like the three dimensions of space, Minkowski's formulation of Special Relativity might at first suggest that it is possible to travel through time. If that is true, could the relativity of time destroy the notion of causality? Can causes and their effects be reversed? The answer, according to Special Relativity at least, is a reassuring 'no'. We are each travelling at the speed of light along our own personal time dimension – and nothing can go travel faster than the speed of light.

Is relativity real? Special Relativity is not only rigorous theoretically, but its predicted effects have been verified many times in practice. For example, increasingly sophisticated versions of Michelson's experiment have been used to test the main postulate of Special Relativity: that the speed of light does not depend upon relative motion. All are in agreement with the original finding. An experiment similar to the Michelson's experiment, carried out in 2001 at the University of Konstanz, in Germany, was the most accurate test yet of the absolute nature of the speed of light. The experiment involved a beam of light bouncing back and forth inside a cavity for over six months. During this time, the Earth revolved half way around its orbit, and changed its direction by 180°. So precisely set up was the experiment that the researchers could be sure that the speed of light relative to space did not change by any more than 1 part in one thousand trillion. If the speed of light did depend on relative motion, this experiment would almost certainly have found it. Accepting the main postulate of Special Relativity as fact makes it necessary to accept all of its theoretical consequences. But, of course, these, too have been tested many times.

A good demonstration of time dilation involves the decay of subatomic particles. Inside particle accelerators, described in Chapter Four, tiny particles travel at close to the speed of light. The effect of time dilation is observed very clearly: the particles survive for much longer before decaying (in the laboratory's frame of reference) when they travel at high speeds than they do when they are at rest. The observed time dilation is in agreement with predictions using the equations of Special Relativity. In the frame of reference in which the particle is stationary, it is the laboratory that moves, with a speed close to the speed of light. The distance through which the laboratory moves before the particle decays in this frame of reference is much shorter due to length contraction. A shorter distance implies a shorter time, and so the time it takes the particle to decay in its own, stationary, frame of reference is the same (shorter) value as observed when it is at rest. If the particle were able to measure time, it would notice that time was travelling more slowly in the laboratory that was rushing past it during its short lifetime. It is also worth noting that no particle has ever been accelerated to a speed greater than the speed of light, however much the power of the accelerator is increased – another prediction of Special Relativity. Similarly, in some particle decays, the particles emit electromagnetic radiation, in the form of gamma rays. In all cases, the measured speed of the gamma rays is the speed of light, not the speed of light plus the speed of the particle.

One of the classic tests of time dilation involved sending clocks carried aboard jet aeroplanes. In 1971, two American research scientists, Joseph Hafele (1933 - 2014) and Richard Keating (1941 - 2006), carried out an experiment in which four atomic clocks were carried in aeroplanes around the world. Two clocks travelled eastwards and two clocks travelled westwards. Before the trip, the clocks were synchronised with further atomic clocks at the U.S. Naval Observatory, which provided a reference frame. All the clocks were compared when the travelling clocks completed their journeys. The details of the experimenters' predictions using Special Relativity are complicated, and they also depend upon General Relativity, which I shall investigate in Chapter Three. Nonetheless, the travelled clocks did indeed show different times from the stay-at-home clocks. The amount of difference in each case was very close to the figure predicted, and within the calculated margin of error for the experimental set-up.

Another prediction of Special Relativity is mass-energy equivalence, predicted by Einstein's famous equation. This mass-energy is liberated in nuclear fusion reactions, such as those at the centres of stars, since the helium produced by the reaction have less mass than the hydrogen from which it formed. When energy is liberated from a system, that system's mass is reduced. The Sun is losing mass at a rate of several million tonnes per second, due to its incredible rate of energy production, for example. On Earth, nuclear reactors and nuclear weapons depend upon this effect, too. The bomb that destroyed the city of Hiroshima, Japan, in 1945 was carrying 60 kilograms of highly enriched uranium, which, upon detonation, underwent a fast nuclear reaction, losing about 1 gram of its mass and liberating a huge amount of energy in a few seconds.

Special Relativity provides great insight into the concepts of space and time. But it says more about what space and time are not than what they actually are. I shall explore what this might mean in terms of our understanding of reality in Chapter Seven. I shall look again at time and causality in Chapter Three and I shall examine the question of empty space again in Chapter Four. The theory of Special Relativity is not called 'special' because physicists are fond of it. Rather, it is because it only applies to special circumstances. Its equations are only designed to deal with situations in which the velocity remains unchanging. If the speed or direction of relative motion change, Special Relativity has little to say. Even the cornerstone of Special Relativity – the non-existence of absolute rest – is in question when there is acceleration, because if you are accelerating, there is suddenly a way to tell that you are not at rest. Inside a spacecraft with no windows travelling at a constant speed past another spacecraft, there is no way of telling whether you are moving or not. There is no sense in the question, since who is to say if it is you or the other spacecraft that is moving? If your rocket engines fire, however, you will feel an acceleration: you will know you are moving. Einstein considered this a challenge, and set out to generalise his theory of relativity. In 1916, he published his theory of General Relativity, whose predictions include black holes, wormholes, and the expansion of the Universe. General Relativity and its strange consequences are the subject of our next chapter.

3: GRAVITY AND WORMHOLES

You are in a spacecraft, far from any stars or planets, with the rocket engines turned off. It makes no sense to ask whether you are moving or standing still, since, as we discovered in Chapter Two, there is no such thing as absolute rest. Let go of an object, and it stays hanging in the air in front of you, motionless in your capsule; shine a beam of light across the capsule, and it travels in a straight line, hitting the opposite wall at exactly the same height. Now turn on the rockets, and they accelerate the spacecraft. Suddenly, you feel as though you are being pulled against the back of your capsule. If you could not hear the rocket engines, you might guess that your capsule was sitting back on Earth, being subjected to a force of gravity. The object you let go of falls towards the back of the capsule, accelerating downwards as it would if pulled by gravity. The only effect described here that may seem unconnected with gravity is that beam of light now bends downwards. As we shall see, light does bend under the influence of gravity. The equivalence between acceleration and gravity is at the heart of Einstein's theory of General Relativity. This bold and elegant theory is an extension of Special Relativity to situations involving gravity; and the bending of light under the influence of gravity is one of its key predictions.

The influence of gravity extends over a very long range: it agglomerates galaxies into vast superclusters hundreds of millions of light years across, for example. So, understanding gravity is clearly very important in figuring out how the Universe works. Solutions to the complicated mathematical equations of General

Relativity provide theorists with predictions of phenomena such as black holes and the expansion of the Universe, discussed in Chapter One. Some of the predictions of General Relativity also have far-reaching philosophical consequences, including the possibility of time travel.

The field of gravitation In 1666, Isaac Newton was hit by an idea: that the force that keeps the Moon in its orbit is the same as the force that makes objects fall downwards here on Earth. In 1679, he set out his Law of Universal Gravitation – and showed how it could explain the motions of projectiles, planets, comets and pendulums – in one of the greatest scientific works ever published: *Philosophiae Naturalis Principia Mathematica*, commonly known as the *Principia*. According to Newton's Law of Gravitation, the gravitational force attracting two objects depends upon the masses of the objects and the distance between them. Newton's Law is an 'inverse square law' with respect to distance. This means, for example, that if you increase the distance between two objects by a factor of 2, the gravitational force between them becomes ¼ the strength (since 4 is the square of 2). Move two objects 10 times as far apart, and the force will have only one-hundredth of its initial strength. One of the many successes of Newton's theory was its explanation of tidal forces. Newton was able to show how the difference in gravitational force across large distances – such as the effect of the Moon's presence as felt at either side of the Earth – gives rise to a strain on large objects. This tidal strain is not enough to damage our planet, but it does cause a bulging of the oceans which gives rise to the tides.

But Newton's wonderful theory had problems: one of the most prominent concerned the orbit of planet Mercury. According to Newton's law of gravitation, the orbit of Mercury should remain stable, so that the furthest point of its orbit – the perihelion – would be in the same point in space each revolution. However, astronomers through the 19th century had noticed a very slight shift in the position of Mercury's perihelion. It was tiny, but it was significant, because it suggested that Newton's theory was flawed.

Newton's theory does not suggest how gravitational force can act across large distances in space. The physicists of the 19th century had an answer: a force field. The concept of fields was a big deal to 19th century physicists, as it is to physicists today. As discussed in Chapter Two, James Clerk Maxwell worked out equations that describe electric and magnetic fields, and was able to show how undulations in the fields transmit electromagnetic waves. For every point around a magnet, there is a corresponding value of the magnetic field. Another magnet placed within the field will interact with it, producing a force. In the middle of the 19th century, Newton's Law of Gravitation was also interpreted in terms of a force field. Around the Sun, for example, is a gravitational field whose strength, or 'potential', is weaker the further you are from the Sun. The force of gravity attracting a particular planet depends upon the mass of the planet and the gravitational potential at its location in the field. Whether fields really exist, or are simply a mathematical convenience, is a philosophical question without a satisfactory answer. Einstein's theory of General Relativity is a field theory of gravity, but it is fundamentally different from Newton's Law of Gravitation. While Newton was at a loss to account for the mechanism behind gravity, Einstein found one.

More importantly, however, Einstein's theory is more accurate than Newton's. General Relativity addresses the findings of Special Relativity, which are at odds with Newton's Theory of Gravitation. Specifically, Newton's theory assumes that the distance between gravitating objects is absolute, which Einstein's Special Relativity shows to be false. And, perhaps more importantly, Newton's theory makes a tacit assumption that gravitational interaction is instantaneous. Even galaxies millions of light years away have a small but real gravitational influence; in Newton's theory, any changes in that gravitational influence – caused by a shift in position of a galaxy, for example, would be felt immediately. This is in conflict with the findings of Special Relativity, which shows that no information or influence can travel faster than light speed without destroying causality. In fact, General Relativity predicts that the 'speed of gravity' is exactly the same as the speed of light. Einstein wanted his theory of General Relativity to be a replacement for Newton's theory of gravitation. To do so, it would have to explain all of the phenomena related to gravitation – including the observed behaviour of stars and planets, and the origin of tidal forces – but would have to be compatible with Special Relativity and account for the observed changes in Mercury's orbit.

The equivalence principle The key to General Relativity is the equivalence between gravity and acceleration. At the root of this equivalence is the concept of mass. Traditionally, there are two definitions of mass: 'gravitational mass' and 'inertial mass'. On Earth, the more (gravitational) mass an object has, the more it is pulled by Earth's gravitational field, and the more it weighs. So, gravitational mass is the '*m*' in the formula describing Newton's Law of Universal Gravitation. Inertia is the resistance to a change in motion: it takes more force to accelerate a heavy (more massive) lorry than a lighter (less massive) car, for example. Inertial mass is the '*m*' used in the formula describing Newton's Laws of Motion. Gravitational mass and inertial mass are very different concepts, and there is no reason why they should have exactly the same value. The fact that they do leads to a strange and counter-intuitive phenomenon: because a heavier object is pulled more strongly by gravity, but also resists that pull more strongly, objects of very different masses accelerate at the same rate through a gravitational field. The most striking demonstration of this fact was carried out on the Moon, by astronaut David Scott (born 1932) during the Apollo 15 mission in 1971. In one hand, Scott held a feather, in the other hand, a hammer. When he let them go, both objects fell together, and landed at exactly the same time. The reason this experiment does not normally work on Earth, of course, is that air resistance slows the descent of the feather.

Scientists were aware of the equivalence of inertial mass and gravitational mass long before the Apollo missions. However, no one could work out why it should be so. It remained a mystery until one day in 1907, when Einstein had what he described as the happiest thought of his life. Einstein realised the importance of a simple fact: that someone falling freely through a gravitational field does not actually feel the force of gravity. In other words, the laws of physics in a freely falling frame of reference are the same as in the 'stationary' frames of reference that feature in Special Relativity. To see why, imagine yourself once again in a spacecraft with the rocket engines turned off, but this time somewhere fairly close to a massive star. Gravity will cause your spacecraft to accelerate towards the star, assuming there is nothing to stop it. Everything aboard the craft, including you, will also accelerate, and at exactly the same rate. Drop an object from your hand and, relative to the spacecraft, it will stay still. Shine a beam of light across the capsule, and it will travel in a straight line. In fact, in such a situation, all the laws of physics would be familiar, and there would be no way for you to tell if you were accelerating under the influence of gravity or if you were stationary in free space (until you eventually plummet into the star, of course). In a very real sense, free fall in a gravitational field really is the same as 'standing still'. If you fire up the rocket engines just enough to stop you moving any closer to the star, you would consider yourself as stationary with respect to the star. But you will feel the same acceleration as you did when you fired the rockets in free space, including the sensation of gravity that we feel on Earth. There will be a directionality to the laws of physics: objects will 'fall', and even light will bend. But this time, you will be accelerating without apparently moving through space. On Earth, the ground exerts a force that acts just like the rocket engines in this scenario: it prevents our free fall towards the centre of the planet. The reason we feel our weight and observe the familiar effects of gravity is that the ground is constantly accelerating us upwards!

Imagine two spacecraft freely falling towards a star, sending each other parcels and light rays. In the frames of reference of each spacecraft, the other is at rest, the star is accelerating towards them, and the light rays and parcels travel in straight lines. From a frame of reference at rest relative to the star, the trajectories of the parcels – and light rays – are curves called parabolas. The frame of reference at a fixed distance from the star is the result of firing the rocket engines just the right amount, making the spacecraft 'hover'. In Newtonian physics, this would be interpreted as the force produced by the rockets balancing the force produced by the gravitational field. In Einsteinian physics, gravity is not seen as a force at all; the acceleration that you would feel in this hovering scenario is the same as you would get if the rocket engines were firing and giving you real acceleration in 'free' space, far away from any stars. The all-important question is "How can accelerating towards a star under the influence of gravity be equivalent to not moving at all, while not moving at all relative to a star feels like acceleration?" The answer concerns the geometry of spacetime.

Einstein's theory of General Relativity allowed him to reconcile Special Relativity and gravity, by finding a way to put all frames of reference on an equal footing. As we shall see, in order to derive the finished theory, Einstein had to extend the concept of the spacetime continuum, introduced in Chapter Two. Along the way to the finished theory, however, two important and interrelated phenomena arose just days after Einstein had his initial happy thought: gravitational time dilation and gravitational redshift. Gravitational time dilation is a phenomenon that causes clocks at different gravitational potentials to run at different rates. The stronger the gravitational potential, the slower a clock will tick relative to someone in free space. As a result of this gravitational time dilation, a clock 10 kilometres above Earth's sea level will gain about 0.00003 seconds in the course of a year relative to a clock at sea level. Similarly, one second passes at the surface of the Sun – at a stronger gravitational potential – for every 0.999998 seconds at the surface of the Earth. This adds up to a difference of 0.00073 seconds over a year. In yet stronger gravitational potentials, the effect is much more significant. For example, a neutron star is a very dense object indeed, so it has a high gravitational potential close to its surface. Many hours pass on Earth for every hour at the surface of a neutron star.

The second phenomenon, gravitational redshift, is not unrelated to gravitational time dilation. To explain gravitational redshift, consider shining a laser beam from the ground up into the air. The frequency of the laser beam is very precise – one of the defining characteristics of a laser. Measure the frequency of the laser beam high in the air, and it will be very slightly lower, and the wavelength very slightly longer, than you measure at ground level. The reason is that the light loses energy in freefall in a gravitational field. A stone thrown upwards will lose kinetic energy, and slow down as a result. But light cannot slow down; the laser beam must lose energy in another way. Since the frequency and wavelength of light are directly related to its energy, they must change. This phenomenon can also be understood in terms of gravitational time dilation: the frequency is defined relative to time (and so is the wavelength, since the speed of light is constant). Since time runs slower at ground level than 10 kilometres above ground level, it makes sense that the frequency will be reduced when measured high in the air. Again, the effect is minuscule, becoming significant only in very strong gravitational fields. In order to formulate his new theory, Einstein needed to account for gravitational time dilation and gravitational redshift. He also had to incorporate the idea that light should bend in a gravitational field. He eventually addressed all of these issues with a radical but inescapable solution: spacetime must be distorted, somehow 'warped' or 'curved'.

Warping spacetime Spacetime is a geometrical representation of the Universe, as a four-dimensional 'fabric'. Every event that has ever taken place or will ever take place is defined by a point in spacetime. The collection of events that constitutes spacetime is ordered and absolute, but measurements of time and space within it are relative. Each observer has a unique, straight-line trajectory through spacetime, called a world line. In your own frame of reference, your world line is aligned along your own time dimension, and time passes at a consistent rate. In Special Relativity, time dilation (and length contraction, both discussed in Chapter Two) can be explained by the fact that observers in relative motion do not share the same time dimension. When Hermann Minkowski – Einstein's old mathematics teacher – reformulated Einstein's Special Relativity in terms of spacetime geometry in 1908 (again, see Chapter Two), Einstein was not impressed. He thought that Minkowski's geometrical interpretation obscured the underlying physics. However, many other physicists quickly recognised Minkowski's idea as an excellent way of representing the effects described by Einstein's theory. Einstein himself became enamoured with the idea of spacetime geometry when he realised that it was not only a good tool for explaining the concepts of Special Relativity, but is essential in the formulation of General Relativity.

In Special Relativity, the world lines of observers that could be considered to be stationary – those in unchanging motion – are straight lines in spacetime. In order to reconcile gravity and Special Relativity, Einstein would have to find a way to retain the straight-line world lines. This is why, in 1912, Einstein proposed that the presence of mass-energy causes spacetime to distort: to curve, or 'warp'. You can get a feel for why curved spacetime causes gravitation by considering two ships travelling due north from the equator on Earth. The Earth's surface is two-dimensional, and is therefore a poor representation of the curvature of spacetime. Nevertheless, you can see that the two ships, travelling parallel to each other, and each apparently in straight lines, move towards each other at an increasing rate. This is similar to what happens as two objects travelling through space time accelerate towards each other. You can also get a feel for how warped spacetime causes tidal forces – an object that is extended over a large volume of space will have different parts in regions of different curvature, resulting in a strain on the object as a whole. So, the ocean tides here on Earth are caused by the warping of spacetime by the presence of the Moon.

Einstein struggled with the mathematical formulation of General Relativity in terms of warped spacetime: "Compared to this problem, the original theory of relativity is child's play", he wrote to a colleague. In warped spacetime, even the orbit of the Earth around the Sun could be considered as a straight line. Einstein wondered if any mathematicians had investigated curved space. With the help of a mathematician friend of his, Marcel Grossman (1878 - 1936), Einstein learned of the work of German mathematician Bernhard Riemann (1826 - 1866). In his dissertation thesis in 1854, Riemann had derived a way of analysing multi-dimensional space. He even raised the question of what the geometry of real space might be like. In his paper, Riemann had derived a 'tensor': an algebraic expression for working out the curvature of space at a particular point.

Riemann's ideas about the geometry of multi-dimensional spaces were to be crucial in turning four-dimensional spacetime into something that could deliver gravity in accordance with Einstein's General Relativity. In 1915, Einstein found a way to equate the density and pressure of mass-energy with Riemann’s curvature tensor. Put simply, mass-energy curves spacetime. The result is the Einstein Field Equations – a set of ‘coupled hyperbolic-elliptic nonlinear partial differential equations’. The mathematics of Einstein's Field Equations is extremely complicated, but we can see in principle how the curvature of spacetime can explain what we think of as gravitational attraction. In the 'flat' spacetime of Special Relativity, standing still involves movement through time, but does not involve movement through space, in your own frame of reference. But in warped spacetime, travelling through time necessarily moves you through space, too. Place yourself in the vicinity of a massive star with no initial speed, and as you move through time – in other words, by just existing – you move through space. You will find yourself drifting towards the star.

Four-dimensional spacetime is difficult enough for us to visualise in Special Relativity, where it is ‘flat’. But the warped spacetime of General Relativity presents us with an even greater challenge. In order to visualise the geometry of spacetime in specific regions, such as around a star, physicists often use an embedding diagram. In such a diagram, the three dimensions of space are represented as a two-dimensional surface, while the curvature of spacetime can be represented as curvature of that surface, into a third dimension. This is the familiar 'rubber sheet' analogy, which is useful, but limited in its relationship to reality. An embedding diagram attempts to illustrate something called hyperspace, which is itself an abstract mathematical simplification of what warped spacetime is really like. The reason it is called an embedding diagram is that it visualises curvature in two dimensions by embedding it in a three-dimensional picture. Spacetime is curved, but it is not curved 'into another dimension'. There is a direct connection between the mathematical co-ordinates of hyperspace and those of 'real' spacetime, so embedding diagrams are invaluable tools for visualising the phenomenon of warped spacetime. The main problem with them is that the time co-ordinate is not included: an embedding diagram is a snapshot in time. For static situations, like the description of the gravitational field around a star, this is useful and relevant.

It is impossible to visualise the curvature of a three-dimensional space, let alone four-dimensional spacetime. However, the curvature of two-dimensional surfaces is familiar and easy to visualise, and is defined in the same way mathematically. The outside of a sphere has a curved surface – by convention called a positive curvature; the inside surface of a sphere or the surface of saddle has a negative curvature; a flat surface, like a floor (and also, mathematically speaking, the surface of a cylinder), has zero curvature. Expanding this visualisation to four dimensions is impossible, but Riemann showed that the same principles apply, and that it is possible for beings to "discover the curvature of their Universe, and compute it at any point". Einstein made use of this idea in his theory of General Relativity. An embedding diagram shows the gravitational field around a star, as a warping of space.

There is an infinite number of solutions to Einstein's field equations for spacetime curvature, each one depending upon the exact distribution of mass-energy. The simplest case involves a spherically-symmetric distribution of mass – a 'ball shape', such as a star. German mathematician Karl Schwarzschild (1873 - 1916) first worked out the solution to Einstein's equations for such an object in 1916. Not surprisingly, Schwarzschild's elegant solution took the same form as the gravitational field that Newton's theory of Gravitation would predict, with the field strength varying as the inverse square of the distance from the centre of the star. It also predicted gravitational time dilation, neatly in line with the principles of General Relativity. For every star, the time dilation at its surface depends upon its mass. The more mass contained within the surface of the star, the more intensely the spacetime is warped at its surface. For each mass of star, there is a corresponding size that would result in an infinite time dilation. This would be true for the Sun if all of its mass were contained within a radius of about 18 kilometres, for example. That crucial radius is called the Schwarzschild radius. The surface of an extremely dense star – one whose mass is contained within its Schwarzschild radius – becomes a barrier through which no information can pass from the star out into the Universe. If the mass is confined within an even smaller radius, the Schwarzschild radius is still significant: time dilation there is still infinite. If time dilation is infinite, gravitational redshift is infinite, and no light or other signal can leave from inside it. A star inside its Schwarzschild radius is hidden from the rest of the Universe inside an 'event horizon'. If you were able to hover above an event horizon, stationary in the extremely warped spacetime there, you would see time in the Universe around you rushing past at an incredible rate. All the electromagnetic radiation that reached you from the Universe would be gravitationally blue-shifted, not redshifted, to a large degree. So, what was visible light would now be X-rays or even gamma rays.

More worrying still for Einstein, Schwarzschild's solution suggested that inside the event horizon, all the star's mass would necessarily be crushed to an infinitesimal size by the very curvature of space that its presence brings about. It would then occupy a point of infinite density called a singularity. Fortunately, the mass of most stars is not concentrated within Schwarzschild's radius. An object with the mass of Earth would have to be compacted to the size of a pea to qualify. But what if there was an object whose mass was contained within the Schwarzschild radius? Would time really slow down to nothing at the critical radius, relative to someone far away in free space? Would electromagnetic radiation leaving the surface really be gravitationally redshifted out of existence? Through Einstein's new theory, Schwarzschild seemed to have suggested the existence of the strange objects mentioned in Chapter One: black holes.

Black holes The concept (but not the name) of black holes was first proposed by English geologist and astronomer John Michell (1724 - 1793), as long ago as 1783. Michell came upon the idea while considering escape velocity: the speed at which an object must be travelling to completely escape the gravitational influence of a massive object. If you shoot a bullet vertically upwards on Earth at, say, one kilometre per second, it will fall back down. If you shoot the same gun vertically on the Moon, the bullet will travel much further up from the surface, but will still come down. But from the surface of a small asteroid, the bullet would never return. Your bullet would have to be travelling at about 11 kilometres per second on Earth to have the same effect; this is the escape velocity of planet Earth. The escape velocity of the Moon is about 2 kilometres per second, while the escape velocity of the Sun is about 41 kilometres per second. Michell calculated what the escape velocity would be for a supermassive star with the same density as the Sun, but 500 times the diameter, and found that it would be greater than the speed of light. In Michell's day, physicists believed that light was made of material particles, which would be attracted by gravitational force just like everything else. And so, reasoned Michell, such a supermassive star would be invisible, and the Universe could contain many of them without our knowledge. Thirteen years later, French mathematician Pierre Laplace (1749 - 1827) independently originated the same ideas. However, during the 19th century, physicists were convinced that light is a wave motion, and therefore assumed not to be affected by gravity. Interest in 'dark stars' waned.

Einstein's theory of General Relativity inadvertently reawakened interest in the strange invisible objects hypothesised by Michell and Laplace, thanks to Schwarzschild's solution. With the new interpretation of gravity in terms of warped spacetime, it made no difference whether light is a wave motion or a stream of particles: it will bend in a gravitational field. Einstein was unwilling to accept the idea that a singularity, a point of infinite density, would have to lurk at the centre of the objects predicted by Schwarzschild. He was sure that nature must have a way of outlawing such a phenomenon. As we saw in Chapter One, nature does indeed have ways of preventing the gravitational collapse of massive stars up to a certain extent. In a main sequence star, thermal pressure as a result of nuclear reactions prop up the star. Electron degeneracy, the pressure that prevents a white dwarf from further collapse, was proposed in the 1920s; neutron degeneracy that does the same for a (denser) neutron star was put forward in the 1930s. But by the end of the 1930s, physicists began to realise that there seems to be nothing beyond neutron degeneracy that can prevent the formation of a singularity. As physicists came to terms with the existence of singularities, they invested time in examining the features that a black hole might possess.

During the 1940s and 1950s, slow but definite progress was made by physicists studying black hole physics. But between 1965 and 1975, black hole research – and the search for real black holes – really took off. This was largely due to the solution of Einstein's equations for a rotating star, a solution put forward in 1963 by New Zealand physicist Roy Kerr (born 1934). Other physicists correctly interpreted Kerr's solution as describing a rotating black hole. This was because the solution made no mention of the internal structure of the star, and black holes have no internal structure. Beyond the distinction that some black holes may rotate at different speeds, there are only two factors that distinguish one black hole from another: mass and electric charge (not including, of course, their position in spacetime). This was summed up by American physicist John Wheeler (1911 - 2008) with the phrase 'black holes have no hair'. Notwithstanding differences in mass, charge and spin, one black hole is exactly like any other. But what would these strange objects really be like?

Contrary to much popular belief, black holes do not suck up everything like a huge cosmic vacuum cleaner: if Earth suddenly became a black hole, it would still orbit the Sun, and the Moon would still orbit it. The International Space Station, too, would be safe, and would continue largely unaffected. You and I, however, would not be so lucky: with the ground no longer present to accelerate us upwards, we would find ourselves in free-fall towards the invisible ex-Earth singularity. Spacetime around the event horizon would be so tightly curved that our bodies would be pulled apart by tidal forces long before we reached it. To the astronauts on the International Space Station, our journey down to the event horizon would take for ever, since time slows down in a strong gravitational field. In our own frame of reference, however, our journey would be all-too-quick.

Many people wonder how black holes can have a gravitational field, given that General Relativity claims that gravity travels at light speed while Special Relativity demands that no information can travel faster than light speed. Surely even gravity should be hidden behind the event horizon of a black hole? This analysis is not correct, however: black holes would indeed have a gravitational field, warping spacetime in just the same way as the star from which they collapsed. The way around the conceptual problem is to bear in mind that the field – the warping of spacetime – around a star is the result of matter just before the star collapsed below the event horizon. The same applies to the electromagnetic field, but in a slightly different way. Electromagnetic fields are produced by electric charges: a charge that does not move is surrounded by an electric field, but when the charge moves it produces a magnetic field, too. The movements of charges cannot be communicated from within the event horizon, so the magnetic field disappears. Only the electric field remains. So, a black hole must have an unchanging electric field, again frozen in time at the moment the collapsing star shrunk within the event horizon. Quantum physics, discussed in Chapter Four, has another perspective on this question, which claims that the electromagnetic field is 'carried' by virtual particles, which can travel faster than light but not carry any information.

Because no electromagnetic radiation can escape from black holes, you might think they are impossible to detect. However, this is not the case: those gravitational fields and, if they are electrically charged, electric fields, can interact with other objects in the Universe, and astronomers should be able to infer from the behaviour of the other objects. In other words, invisible black holes can give themselves away by their effects on visible matter nearby. Fortunately, a large proportion of stars form in pairs (or triplets), co-revolving in a mutual orbit. When one of a pair is massive enough that it eventually forms a black hole when it collapses, its gravitational field pulls gas off the surface of its partner star, as described in Chapter One. On its way 'into' the black hole, the gas forms an accretion disc, and heats up so much that it produces X-rays, giving astronomers a clue as to the presence of the black hole. Many such accretion discs have been observed around massive invisible objects, providing a great deal of evidence of the existence of black holes. However, black holes can be loners, too. And, with a little luck, they too can be discovered, thanks to the phenomenon of gravitational lensing, which relies upon the fact that gravitational fields bend light. Light from a distant object deviates as it passes by a very massive object en route to Earth, in a similar way to light bending as it passes through a lens. If there is a black hole between Earth and a distant galaxy, then astronomers should be able to observe this lensing effect. Several strong candidate lone black holes have been discovered in this way.

In the early 1970s, the idea of black holes with 'no hair' apart from mass, charge and rotation was dealt a blow, largely by the work of British physicist Stephen Hawking (born 1942) and Israeli physicist Jacob Bekenstein (1947 - 2015). Between them, Hawking and Bekenstein realised that black holes possess entropy. To see why, consider what would happen if two black holes collided and combined. This is what Hawking did mathematically in 1970. He realised, with the help of some elementary algebra, that if two black holes combined the event horizon of the new black hole would always have a greater area than the event horizons of the two black holes from which it formed put together. Innocent and unsurprising perhaps, but it was to be very important, for it reminded Hawking of entropy: always on the increase. We encountered the concept of entropy in Chapter Two: it is a measure of disorder or randomness. But how can a black hole have entropy? Surely the concept of entropy is not relevant to a black hole: what can be random about an unchanging mass, unchanging electric field, and a definite rotation? Enter Bekenstein. He reasoned that the area of the event horizon around a black hole might itself somehow represent the black hole's entropy. At first, Hawking was unconvinced. But Bekenstein considered what would happen when matter crosses the event horizon and is lost to the hole. Matter has entropy; would this entropy be lost from the Universe in the no-entropy black hole? If so, then the entropy of the Universe would be reduced, in conflict with the second law of thermodynamics discussed in Chapter Two. But thermodynamics is concerned with heat and temperature; if a black hole has entropy, can it have a temperature? And how might this relate to General Relativity, which suggests that a black hole is nothing but a singularity hidden from the Universe behind its event horizon? Hawking was still not satisfied.

Bekenstein persevered, and he managed to find a way to relate a black hole's entropy to the surface area of its event horizon. Hawking was fairly convinced by this argument. However, there remained another problem. The laws of thermodynamics also demand that any object with a temperature above absolute zero – in other words, any object with any temperature at all – should emit electromagnetic radiation. The spectrum of this thermal radiation should be characteristic of temperature; this explains why the colour of a hot object changes from red, to yellow, to white as you increase its temperature. However, according to General Relativity, a black hole cannot emit any radiation at all – the event horizon puts paid to that. Nonetheless, in 1974, Hawking was able to show that black holes really might emit radiation. He did this by combining the picture of black holes presented by General Relativity with the strange findings of another pillar of modern science, quantum physics. I shall explore the world of quantum physics, and its role in Hawking's black hole radiation, in Chapter Four. Importantly, Hawking found a way to work out a black hole's temperature, and found that it is directly related to its mass. The more massive a black hole, the lower its temperature. This enabled him to determine the nature of the radiation a black hole should emit. And he found that it would indeed be thermal radiation, characteristic of temperature, just as thermodynamics requires. Black holes, it seems, can emit electromagnetic radiation, and they obey the laws of thermodynamics, just like ordinary matter does.

According to Hawking's calculations, the temperature of a typical, stellar black hole would be very low, so the radiation it emits will be very weak. But the fact that the calculations predict that a black hole should emit thermal radiation at all is exciting and challenging. Hawking radiation, as it has been called, would certainly be swamped by the powerful X-rays emitted by the hot accretion disc of a typical black hole. It would even be weaker than the cosmic background radiation we receive from every direction of space, described in Chapter One. It comes as no surprise, then, that Hawking's radiation has yet to be observed. However, most physicists are convinced that it must exist.

Assuming that black holes radiate, they must also lose mass, by virtue of the mass-energy equivalence derived in Special Relativity. So, a massive, cold black hole would slowly 'evaporate', losing mass as it radiates. As it loses mass, its temperature must rise, since a less massive black hole has a higher temperature. And so, as time goes on, the black hole will radiate energy, and lose mass, at an increasing rate. Eventually, when the mass is reduced to perhaps a million tonnes, with an event horizon much smaller than the size of an atom, the black hole would be unbelievably hot, and its rate of emission of radiation would be so great that it would violently explode. Based on these ideas, the lifetime of a stellar black hole would be billions and billions of times the estimated age of the Universe (which I will discuss in Chapter Six). However, a black hole that starts tiny would not last long at all. In experiments here on Earth, tiny black holes may be created fleetingly in powerful particle accelerators, each one destined for a brief existence concluding with a brief flash of high energy radiation. No such events have been observed yet.

A typical stellar black hole loses mass-energy extremely slowly. Its mass-energy is more likely to increase overall than to decrease, as a result of matter and radiation falling into it. The lifetime of a monster black hole – the supermassive black holes found at the centre of most galaxies, for example – would be longer still, and the temperature extremely low. If you find it challenging that these strange ideas about black holes can follow from Einstein's theory of General Relativity, aided and abetted here and there by quantum physics, then hold on to your conception of reality, because things become more strange and more challenging still from here on.

Tunnels through spacetime At the end of Chapter Two, I considered the possibility of travelling freely through time. Travelling backwards in time is prohibited by Special Relativity, because nothing can travel faster than the speed of light. However, in the new warped landscape of reality formulated by General Relativity, spacetime seems more 'flexible'. In 1949, German mathematician Kurt Gödel (1906 - 1978) highlighted the possibility of world lines that return to the same point in spacetime. These trajectories in spacetime are facets of genuine solutions to Einstein's field equations, and are therefore permitted according to General Relativity. However, a Universe with Gödel's spacetime loops – what have become known as 'closed time-like curves' – is a very specific solution, and is almost certainly not what the real Universe is like. But might there be other solutions to Einstein's equations that describe situations in which it is possible to travel into the far future or the distant past?

Actually, travelling into the far future is quite simple, in theory at least. All you have to do is visit a black hole, wait just above its event horizon for a short time, while time in the rest of the Universe speeds onwards at an incredibly quickened rate, relative to you. Come back to Earth, and you will find it in its far future. Of course, you would have to survive the crushing and stretching caused by the intense curvature of spacetime around the black hole, but if you stay at a safer height over the black hole, you will still achieve your goal – you will just have to wait there a little longer. This scenario does not challenge the concept of causality, and nothing in General Relativity or any other scientific theory prevent it. But what about travelling into the distant past? Clearly, in that case, potential paradoxes arise: for example, you could travel into the past, murder your own grandfather when he was a young boy, and then travel back to the 'present'. Affecting the past clearly affects the present. In this case, you make your own existence an impossibility! But this paradox goes deeper than that: for if you never existed, who killed your grandfather? Despite paradoxes like this, by examining possible solutions to the equations of General Relativity, theoretical physicists have shown that travel into the past may actually be possible … via tunnels in hyperspace.

As long ago as 1916, physicists looking at Karl Schwarzschild's solution to Einstein's field equations realised that if the spacetime curvature at the event horizon of a black hole is infinite, then it could be 'open' at the 'bottom'. In fact, the full version of Schwarzschild's solution naturally includes a surprising symmetry: a white hole, which is the reverse of a black hole, pushing outwards into another Universe, or perhaps into another region of spacetime in our Universe. Warped spacetime would form a continuous connection between the black hole and the white hole. This connection can be visualised in an embedding diagram as a tunnel, and is called an Einstein-Rosen Bridge. Clearly, if you could travel through an Einstein-Rosen Bridge (you would have to survive the crushing spacetime warp along the way) you would be thrown out of the white hole at a very different point in spacetime, and possibly in another Universe. Unfortunately, this perfect, symmetrical situation is part of an idealised solution, in which no other matter or energy present. If you tried to send matter or energy through the bridge, the bridge itself would collapse. The only way to send matter across the bridge would be to push it faster than the speed of light, so that it makes it through the bridge just before the collapse. However, faster-than-light travel is not possible, even in an idealised Universe.

In 1955, John Wheeler coined a more catchy name for these spacetime tunnels: 'wormholes'. In 1963, the concept of using wormholes to connect distant regions of spacetime was given a boost, by Roy Kerr's solution of Einstein's equations, the one which describes a rotating black hole. When a black hole rotates the singularity becomes extended: it is no longer an infinitesimal point, but instead forms a doughnut-shaped ring. Passing through this ring would, in principle, result in travel across spacetime, so that, for example, travelling just a few metres through the ring would be a hyperspace hop that would take you, perhaps, to another galaxy or another time. However, by the 1970s, it had been shown that even a rotating black hole, with its extended singularity, would be unstable. And even if you could travel through the wormhole of a rotating black hole, there would be no way back, since you would have to cross the event horizon in the reverse direction. And the only way to do that is to travel faster than light. But the idea of wormholes was not dead.

In the 1980s, planetary scientist and author Carl Sagan (1934 - 1996) was finishing off 'Contact', a science fiction novel that he was determined would involve as much science fact as possible. Sagan's story featured an intelligent, extraterrestrial civilisation that sends a message to Earth telling them how to build a machine that would enable them to travel immense distances in space. Originally, the machine in the story made a black hole through which a human astronaut would travel across space. But Sagan was unsure whether this would work, so he asked his friend, the American physicist Kip Thorne (born 1940), for help. Thorne knew that a black hole would be no good: its singularity would crush the astronaut. A wormhole, he told Sagan in 1985, is the answer; if only he could find a way for it to stay open. If the astronaut was to travel vast distances – to another star or another galaxy – then spacetime would have to be curved accordingly. If it was, then the locations would be close together in hyperspace, while remaining far apart through 'ordinary spacetime'. In this way, a wormhole could be just the spacetime shortcut Sagan was looking for. The required wormhole would not have a singularity or event horizon, and would therefore enable two-way travel.

Thorne found a possible solution to the problem of keeping an unstable wormhole open: thread it with some sort of matter that can produce an anti-gravity effect, pushing outwards from within the wormhole. The wormhole would not necessarily be a result of the collapse of a massive star: it could be entirely 'engineered' from spacetime. If the matter in question is to push the wormhole apart, it must have negative energy. General Relativity shows how increasing the density of mass-energy in space increases the curvature of spacetime. So, a negative mass-energy would decrease it. This is exactly the desired effect inside the wormhole. In 1987, Thorne and his colleagues put forward an idea of how such negative energy might be harnessed, but also stressed that it may also be possible in principle to keep a wormhole open in other ways. Thorne is one of the biggest proponents of wormhole theory. In 1986, he realised that wormholes would not only allow a civilisation to traverse huge, intergalactic distances instantaneously, but that, as a result of travelling so far in so little time, they would effectively allow travel at speeds faster than light. Wormholes could take you back in time. And in 2000, Russian astrophysicist Serguei Krasnikov (born 1961) showed how it might be possible for a wormhole to create its own exotic matter, sustaining it indefinitely and to any diameter. Perhaps such a wormhole would even be big enough and stable enough for a human traveller to pass through it.

However fanciful all of this might seem, it does suggest that time travel into the past may not be impossible after all, at least in principle. So what about the paradoxes that time travel raises? Physicists tend to assume that something in the laws of Nature must outlaw it. Stephen Hawking calls it the

Chronology Protection Conjecture. It may be that any wormhole would be short-lived after all, exploding with showers of high-energy particles, for example. It may be that all of the events of the Universe are already determined, and that you cannot go back and kill your grandfather because you already did not. This 'consistent histories' approach relies on the supposition that we humans have no free will, which might well be the case. Another possible way out of time travel paradoxes is to assume that our Universe is one of many, 'parallel' Universes, which together form a multiverse. According to this picture, when you travel back in time to kill your grandfather, you travel to a different Universe. In that parallel Universe, the prevention of your birth would not be a problem (and nor, we assume, would the sudden appearance of a time traveller), so you can murder your grandfather without breaking the laws of Nature, at least.

Hawking believes that time travel into the past must be against the Universe's laws. The fact that no one has visited us from the far future is enough evidence to convince him of it. However, in some of the formulations of time machines based on wormholes, it is possible to travel back in time only as far as the point when the worm hole was created. With this in mind, it is no surprise that we see no visitors from the far future. Once we open the first international wormhole spacetime port, we may be inundated with time travellers, perhaps bent on finding their grandparents.

So it appears that spacetime is a flexible 'fabric' of reality, made for moulding. If our descendants are to have the ability to manipulate spacetime at will, they would need access to nearly limitless supplies of energy, and would need to overcome enormous practical challenges. With this mastery over spacetime, intelligent beings elsewhere could already have found ways to cross immense distances in practically no time, just as Carl Sagan envisaged. But if we Earthlings are to be the instigators of such a scheme, then the first step along the road is to make sure that General Relativity is the ultimate theory of gravity.

Testing General Relativity One of Einstein's imperatives when he was formulating his new theory was to explain the shift in Mercury's orbit. As I mentioned near the beginning of this chapter, 19th century astronomers had noticed that Mercury did not behave according to Newton's Theory of Universal Gravitation. Desperate to cling on to the theory that had served them for more than two hundred years, they suggested many explanations of Mercury's behaviour, including postulating the existence of an undiscovered planet that was tugging on Mercury. When Einstein found the final form of General Relativity, he was relieved to find that it did predict the correct shift in Mercury's orbit. The gradual shift in Mercury's orbit is due to the way in which spacetime is warped by the mass of the Sun. However, Einstein's new theory was such a departure from Newton's that physicists and astronomers would need something more substantial to convince them.

The classic test of General Relativity, and the one that would make Einstein an instant worldwide celebrity, was carried out in 1919, during a solar eclipse. Einstein's theory predicted that the light from distant stars would be bent as it passed near to the Sun: very weak gravitational lensing. This would cause the apparent position of the star to shift very slightly as seen from telescopes on Earth. However, the Sun's light normally drowns out the light from stars that lie close to it in the sky, making this effect impossible to observe in daylight. During a solar eclipse, however, the Sun's bright disk is obscured by the Moon, and the sky is darkened. The crucial stars, whose positions were known with great accuracy, did indeed appear shifted very slightly relative to their normal positions. The Sun's mass really was warping spacetime, so that the stars' light was bending as it passed by. The angle by which the stars shifted was exactly as predicted by the new theory (well, convincingly within the range of errors inherent in the experiment).

Two other key predictions of General Relativity are gravitational time dilation and the related phenomenon of gravitational redshift. In 1925, American astronomer Walter

Adams (1876 - 1956) analysed the spectrum of light that had left Sirius B, a white dwarf star. According to General Relativity, the light leaving the strongly warped spacetime at the surface of a white dwarf is redshifted 30 times as much as light leaving the Sun. Adams detected a shift in frequency that was in excellent agreement with what Einstein's theory predicted, given the size and mass of the dead star. In the 1960s, the same measurement was made on light emitted by the Sun, again in agreement with predictions. In the most famous experiment on gravitational redshift, American physicists Robert Pound (1919 - 2010) and Glen Rebka (1931 - 2015) detected the gravitational redshift at Earth's surface. They sent a beam of gamma rays up a tower just 22.4 metres tall, travelling up through Earth's gravitational field (very slightly warped spacetime). The predicted redshift was equivalent to two parts in a thousand trillion ($2{:}10^{15}$), equivalent to just 0.0000000000002 per cent. But Pound and Rebka found it. Repeating the experiment, but this time sending gamma rays down the tower, Pound and Rebka found that the gamma rays were now blue-shifted by the same amount. A similar experiment, carried out in 1976, was Gravity Probe A, which involved firing a rocket up to 10 kilometres. The rocket was carrying a maser (like a laser, but emitting microwaves rather than light). At the top of its flight, the rocket and the maser were momentarily at rest relative to the ground. At that moment, this experiment was like Pound and Rebka's experiment, but with a greater difference in height. Like all of the observations of gravitational redshift, this result can also be interpreted as due to gravitational time dilation. The results of all of these complex experiments, and many more, match the predictions made using General Relativity.

Another major prediction of General Relativity, first observed in 2016, is the phenomenon of gravitational waves. According to General Relativity, disturbances in spacetime can propagate through it, just like ripples on the surface of a pond or sound waves through the air. Like any type of wave, gravitational waves carry energy away from the objects that produce them. The detection of gravitational waves was generally considered as the last great test of the validity of General Relativity. Starting in the 1970s, physicists have devised several ingenious ways to look for them. All of them make use of the fact that should they exist, gravitational waves must pass Earth. As they do so, they should repeatedly push objects very slightly closer together and then stretch them very slightly apart as they distort spacetime locally. The first gravitational wave detectors were metal bars or cylinders. The hope was to detect the tiny strains placed upon the detectors by passing gravitational waves. It was hoped that the detectors would be sensitive enough to discern waves from very energetic events, such as supernovas or the collapse of neutron stars to form black holes. However, even these waves would be very weak by the time they reach Earth. These detectors were not sensitive enough.

More recent attempts to detect gravitational waves have involved interferometers, which can measure extremely tiny shifts in distance. In an interferometer, identical laser beams of are sent in two directions at right angles to each other. They bounce off mirrors some distance away, and so are reunited. At the point where the reflected beams meet, a partially-silvered mirror splits off some of the light from each beam and combines them. Any change in the lengths of the light's paths through the experiment should be detected as interference patterns, light and dark fringes produced by the interaction of the two laser beams. The largest gravity wave interferometer is the Laser Interferometer Gravitational Observatory (LIGO) – and it was here that scientists made the first ever confirmed observation of gravitational waves.

The LIGO experiment is carried out at two widely separated sites, so that any vibrations that result from local disturbances can be ruled out of the experimental results. Each 'arm' of the interferometer is four kilometres long. Even over a distance of four kilometres, a typical gravitational wave only results in changes of distance of around one ten-thousand-trillionth of a metre (10^{-16}m); much smaller than an atom. The LIGO experiment detected two different gravitational waves in 2016 – each was produced by a merging of very distant black holes. The emerging field of 'gravitational astronomy' may provide the only way to 'see' dark matter, and thereby help resolve the issue of missing mass described in Chapter One.

One of the most interesting, if indirect, pieces of evidence supporting General Relativity concerns the study of a distant pulsar in a mutual orbit with another star. As described in Chapter One, a pulsar is a rapidly rotating neutron star that emits very regular bursts of radio waves. Astronomers can use the regular pulses as if it were the ticking of a distant, highly accurate clock in the pulsar's frame of reference. The American astronomers Russell Hulse (born 1950) and Joseph Taylor (1941), who discovered the co-revolving pair of stars, determined that the pulsar completes each orbit in less than eight hours. This means the orbit is very tight indeed, and the stars are close together. It also means that the warping of spacetime must be very strong there. As it co-revolves with the other star, the pulsar passes through regions of spacetime with varying curvature. From Earth, this can be observed as slight, regular variations in the timing of the radio pulses; they are emitted at regular time intervals in the pulsar's frame of reference, but the pulsar's time passes at varying rates as seen from Earth. Two massive stars in a mutual orbit should also emit regular gravitational waves. These waves should carry energy away from the binary system, reducing the duration of each orbit. The binary pulsar system has been observed carefully since 1974, and both of these effects have been observed, in very close agreement with theory.

A phenomenon that is related to gravitational waves is frame dragging. General Relativity predicts that spacetime around a massive, spinning object will be dragged around, like air around a spinning tennis ball. One effect of frame dragging is that the rotating object slows down very gradually. The frame dragging produces gravitational waves, which is the mechanism by which the energy is carried away. In 1997, behaviour that is consistent with frame dragging was observed in the accretion disc of a suspected black hole. And in 2004, a spacecraft was launched into an orbit around the Earth. The satellite is part of an experiment called Gravity Probe B, which aims to detect the frame dragging caused by Earth's rotation.

After a year-and-a-half in orbit and five years of analysis of the data gathered, the project was a success: frame dragging was detected at the rate predicted by Einstein's theory.

To most physicists, the theory of General Relativity is as beautiful, robust and well-tested as it is philosophically challenging and enlightening. However, there are those who believe that General Relativity is fundamentally flawed: that the model of reality it demands is misguided. But General Relativity does not proscribe the nature of reality. It is nothing more than a mathematical framework that accurately describes the rules governing real mass and energy in real space and time. All General Relativity can say about reality is that, whatever it is, it seems to work according to those rules. Nonetheless, theories like General Relativity can help to point theorists in new directions. Applying General Relativity to the whole of spacetime, for example, has led to modern theories regarding the origin and destiny of our Universe. I shall explore some of these ideas in Chapter Six.

Despite all the successes of General Relativity, even physicists who are comfortable with it are faced with a nagging problem. General Relativity is in conflict with the theory that governs the behaviour of matter on a small scale: quantum theory. Quantum theory tells us that the Universe is 'lumpy', made of discrete units of matter and energy, while General Relativity describes a smooth, continuous Universe. Unfortunately, for all its beauty and its genius, General Relativity only applies to macroscopic objects: planets, stars, galaxies, black holes and, should they exist, wormholes. It is an approximation to some more complete, ultimate theory of gravity – one that includes the effects of the microscopic situations described by quantum physics. The quest to unite these two theories, to form a quantum theory of gravity, is the main challenge facing modern physics today. I shall examine the state of this quest in Chapter Five. The theory of quantum physics are as radical and as well-tested as General Relativity –perhaps more so. So, in Chapter Four, I will explore the strange world of matter in the quantum realm.

4: MATTER AND ENERGY

"You cannot see atoms, even with the most powerful microscopes." I remember being struck by this statement when I read it in a book as a child. I also recall the disappointment and confusion it brought me. I realised why scientists were convinced that atoms are real: the existence of atoms explains how matter behaves. The existence of the atom seemed as certain as my own, but the fact remained that they had never actually been seen.

Nowadays, however, experimental physicists can produce beautiful images that are direct representations of atoms at surfaces; for me, this is as good as a photograph. They can create atomic-scale sculptures, and trap individual atoms using laser beams. They can even track directly the paths of particles far smaller and more esoteric than atoms. These achievements are largely due to the developments in a field of science called quantum physics. The theories of quantum physics are the most well-tested and all-encompassing of any scientific theory. But they challenge our common sense notions of reality. And so, in this chapter, I will investigate the weird facts of life in the quantum world.

Something for nothing? You may remember from Chapter Three the suggestion by Stephen Hawking that black holes emit radiation, and that they lose mass as they do so. You may also remember that General Relativity predicts that nothing can escape from a black hole. At face value, these ideas seem contradictory. The resolution of the apparent paradox involves not the release of radiation from within the black hole's event horizon, but the creation of particles – from nothing – in the space just outside it. It is these particles that constitute Hawking's radiation. According to quantum theory, particles are produced in pairs, continuously but fleetingly throughout spacetime. Normally, these particle pairs annihilate each other within a tiny fraction of a second. But in the strongly warped spacetime near a black hole, one member of the particle pair can be pulled into the black hole, while the other travels away into the Universe at large. The particle that ends up in the black hole has 'negative energy', while the other has positive energy. It is as if the black hole is stealing negative energy from space. The net result is that the black hole loses energy and, as required by Einstein's mass-energy equivalence, also loses mass. The particle pairs that are created from nothing cannot normally be observed directly. In a sense, they do not exist at all. And yet, they are real phenomena, arising from force fields that permeate space. Physicists call them 'virtual particle pairs'. An input of enough energy, in this case from the black hole, can make these particles real. In order to understand what a virtual particle is, we need to look at what matter is made of, and the nature of forces that act on it. The story begins with atomic theory and electric and magnetic forces.

We are all familiar with the forces of electricity and magnetism. Rub an inflated balloon on a woollen jumper, and both objects become 'electrically charged'. Bring the charged balloon close to the jumper, but not touching it, and you will feel a small force pulling the balloon and jumper together. Two balloons that you have charged in this way will push apart – again, without even touching. Similarly, hold two magnets close together and you will feel them pulling together or pushing apart. The central mystery is how forces like these can be transmitted through the air, and even across empty space. When Michael Faraday was investigating these forces in the 1830s, he found that they were transmitted along curved lines. At first, he proposed that these lines simply indicate how the particles of the surrounding matter, such as the air, line up – just as tiny iron filings could be seen to line up in a magnetic field. In this way, they could transmit the force, from one particle to the next, between charged or magnetised objects. Like most scientists of the day, Faraday believed that matter existed as solid particles in empty space, as discussed in Chapter Two. So, do the particles of the air pass on the forces through mutual contact? Faraday supposed not. Instead, he suggested that the forces could be passed on from each particle to the next across the gaps between them – but, of course, this still involves forces acting mysteriously across distances in empty space. Faraday was at a loss to explain how that might happen. In 1844, however, he found a way. He proposed that the conventional atomic theory, the idea of matter particles existing in a void, should be scrapped. He replaced it with a very different notion. Matter itself, he reasoned, must be a manifestation of force fields; "substance consists of the powers", as Faraday put it. In this new picture, force at a distance was no longer a mystery, since atoms now existed as 'centres of force' in the continuous fields, rather than solid balls surrounded by empty space. According to Faraday, fields are the only reality. As we shall shortly see, this is remarkably similar to modern physicists' conception of matter.

Faraday coined the term 'field' in 1845. And in 1846, as a result of countless experiments with light in electric and magnetic fields, he suggested that light itself might be the transmission of vibrations of the field. Through experiments with electromagnets and induction coils, Faraday had shown that electricity and magnetism are inexorably linked. In 1873, James Clerk Maxwell managed to 'unify' them in a consistent, mathematical way, and he found that oscillations of electromagnetic fields must indeed travel as waves at the speed of light. In 1887, German physicist Heinrich Hertz (1857 - 1894) discovered radio waves: electromagnetic radiation with frequencies much lower, and wavelengths much longer, than light waves. This was another great success for Maxwell's theory. Maxwell's mathematisation of the electromagnetic field allowed him to side-step explanations of the physical nature of fields: the numbers made sense, whatever the physical reality. However, Maxwell and other prominent scientists of the time did propose physical mechanisms to explain fields. Most of them referred to the existence of the aether, either as a continuous fluid or made of solid particles. Within some of these theories was the concept of charged particles called electrons, which were supposed to exist either in the aether or as constituents of matter. Faraday's idea of atoms as fuzzy manifestations of the force field, rather than solid balls, was abandoned, partly because it is an uncomfortable one. It is not as disconcerting to think of space, force and matter as truly tangible things as it is thinking of space as imponderable nothingness and matter and force as invisible fields; such was the lure of the 'mechanistic' view of the world.

Neat and satisfying though this mechanistic interpretation of reality may be, several developments around the end of the 19th century showed that it is fundamentally flawed. We saw in Chapter Two that Einstein's theory of Special Relativity, published in 1905, denied the existence of the aether by outlawing the absolute nature of space and time. In the same year, Einstein published another landmark paper, which was to challenge the physicists' cosy interpretation of reality even more dramatically.

Chunks of light Einstein's paper – *Concerning an Heuristic Point of View Toward the Emission and Transformation of Light* – weighed up Maxwell's electromagnetic theory in the light of two puzzling observations. In each case, theory and experiment were in fundamental disagreement. Naturally, the experiments had to be right, as long as they were carried out carefully and were repeated; you cannot argue with Nature. So the theories were flawed. And yet they were based on Maxwell's rigorous and highly successful theory of electromagnetism. Despite the best efforts of the physicists of the day, theory and experiment remained stubbornly at odds with each other.

The first phenomenon that Einstein's paper tackled was thermal radiation: emitted as a result of atoms vibrating. As atoms vibrate, the charges they contain also vibrate, and that causes a disturbance of the electromagnetic field. They emit radiation, which has the same frequency as the frequency of their vibration. The range of frequencies at which particles vibrate, and the total energy of vibration, depend upon temperature. That is why a graph of the intensity of thermal radiation versus its frequency is characteristic of temperature. This explains how astronomers are able to know the temperatures of distant stars. It was the inability of Maxwell's theory to reproduce the correct shape of the graph that troubled physicists. In 1900, German physicist Max Planck (1858 - 1947) reinvestigated the profile of the radiation emitted by hot objects. Planck applied a new mathematical analysis, assuming that the particles of matter could only vibrate at certain 'allowed' energies, not just any energy. In scientific terms, what he was proposing is that vibrational energy levels might be discrete, not continuous. To use a more pertinent term, the energy of a vibrating electric charge is quantised. Each vibrating atom would emit its energy in tiny chunks that are proportional to the frequency of vibration. So, an atom vibrating very rapidly could only emit relatively large chunks of energy, while an atom vibrating less rapidly could emit smaller chunks. To Planck, this was a mathematical exercise, but it did yield a prediction of the radiation spectrum that was in complete agreement with the experiments.

It took the insight of Einstein to interpret Planck's idea of the restrictions on allowed energies as a physical fact, not just a mathematical curiosity. But Einstein went further: he showed that not only is the vibrational energy of particles discontinuous, but electromagnetic radiation must be emitted in discrete bursts of energy, which he called photons. Each photon is a 'quantum' of the electromagnetic field. For a particular frequency of radiation, each photon has the same energy; the higher the frequency, the more energy each photon possesses. A single photon of visible light, then, has more energy than a single photon of (lower frequency) radio waves.

In proposing this idea, Einstein was able to explain the puzzling result of the second phenomenon on which he focused his attention: the photoelectric effect, first observed by Hertz. Shining light onto a metal surface made the metal release negative charge. The more intense the light, the more charge was released. The electric charge was correctly interpreted as being carried by tiny particles, electrons, that were initially part of the metal. It was known by then that matter was somehow made up of positive and negative charges, so the electrons were assumed to be part of each atom. And so, the electrons are initially bound to the metal – otherwise they would leak out even when not illuminated – and the electromagnetic radiation was clearly supplying the energy necessary to eject them. There were two puzzling aspects of the photoelectric effect. First, below a certain frequency of incident light, no electrons were emitted from the metal at all, however intense the light. Second, above that threshold frequency, even very dim light would cause some electrons to be emitted. By thinking of light as a stream of individual photons, the effect made sense. Above a particular frequency of light, each photon would have enough energy to eject an electron from the metal. The more intense the light, the more photons were supplied each second, and the more rapidly the electrons could be released. But below that threshold frequency, no electrons can ever be ejected however intense the light, because not one of the photons will have enough energy to do so.

In Planck's explanation of thermal radiation, the energy was emitted in discrete chunks, but that did not mean that the chunks are necessarily particles. They could be bursts of radiation with just the right amount of energy, travelling out in all directions. In the photoelectric effect, however, individual photons seem to be ejecting individual electrons. This suggests strongly that photons are localised, like 'particles'; if the photon were spread out in space, an individual electron could not absorb all of its energy. However, throughout the 19th century, even before Maxwell constructed his theory, it had been assumed that light was a wave, and with good reason. Light 'diffracts': it spreads out as it meets a corner or as it passes through tiny gaps, just as water waves splay out around a harbour wall. Light also 'interferes': two light waves falling onto the same surface produce dark and light patches, where the energy of the vibrations cancel out in some places and reinforce in others. Again, this can be observed in water: where there are two sources of vibration in the same body of water, the surface can become doubly choppy in some places, and completely calm in others. Diffraction and interference are wave behaviours; they are hard to explain in terms of particles. And, of course, Maxwell's theory of electromagnetism had produced a wave equation that predicted a speed exactly matching the speed of light.

One observation that sits happily in both interpretations of electromagnetic radiation, as particles and waves, is the fact that light exerts pressure. In 1900, Russian physicist Pyotr Lebedev (1866 - 1912) carried out carefully controlled experiments, and succeeded in measuring the pressure that light exerts on a surface. Lebedev was attempting to find support for Maxwell's electromagnetic wave theory, which predicts that electromagnetic radiation carries momentum. It is difficult to visualise electric fields carrying momentum, but it is completely consistent with Maxwell's wave theory. On the other hand, it is easy to visualise light as carrying momentum and exerting pressure if it is made of particles. And Einstein had found out how to calculate the energy, and therefore the momentum, of an individual photon. They are directly related to its frequency.

Support for the emerging picture of a quantised Universe came in 1913, when Danish physicist Niels Bohr (1885 - 1962) applied it to the structure of the atom. Two years earlier, New Zealand born British physicist Ernest Rutherford (1871 - 1937) had worked out that every atom has its positive electric charge contained within a tiny volume at its centre, the nucleus, around which the electrons seem to be in orbit. Just as planets of the Solar System are held in orbit around the Sun by gravitation, Rutherford believed that electromagnetic forces between the negatively-charged electrons and the positively-charged nucleus would keep the electrons in orbit. This idea seemed to be on the right track, but there was no way to determine the size of an electron's orbit. There was another problem: an electron in orbit is continuously changing direction. This change of direction is an acceleration, and an accelerating electric charge should, according to Maxwell's theory, emit radiation continuously. This emission of radiation would result in a loss of energy from the electron, which in turn would result in the electron spiralling inwards, ever closer to the nucleus in a fraction of a second. Bohr's way around this was to apply the ideas of Planck and Einstein, and suggest that the electrons' levels of energy in the atom were quantised: an electron is only allowed certain energies. Bohr realised what this meant: electrons could gain or lose energy only in discrete amounts, between the allowed energy levels. Electrons do not spiral in towards the nucleus because, to do so they would have to lose energy continuously.

Bohr realised something else. For an electron to change from one allowed energy to another, it would have to gain or lose a certain amount of energy. When an electron loses energy, it might produce a photon, and when it gains energy, that energy might be supplied by a photon, he surmised. If the atom's energy levels are quantised, the photons will only have certain allowed frequencies, too. Bohr knew that atoms do indeed only emit and absorb light at certain frequencies. Chemical elements can be identified by the spectrum of the light they emit – that is how astronomers know what elements are present in distant stars. Bohr was able to test his idea by looking at the spectrum of the simplest atom, hydrogen. He worked out what the energy of a photon would be for each frequency that appears in the hydrogen spectrum, and could then work out the energy levels of the hydrogen atom. When he did, he found that a quantity called angular momentum increases by exactly the same amount each time the electron moves from one energy level to the next. The angular momentum of an electron depends on its mass, its orbital speed and its distance from the nucleus. The fact that it can only increase by a certain amount was a triumph for Bohr's explanation of electron orbits and for the emerging quantum picture of the world. Bohr's idea heightened the debate about the nature of photons. Here, just as in the photoelectric effect, it seemed that photons were behaving like particles. For when an electron absorbs the energy of a photon, all of that energy must be in one place. But if light consists of a stream of particles, why does light behave so much like a wave?

Making waves In 1923, French physicist Louis de Broglie (1892 - 1987) made an inspired leap of reasoning concerning particles and waves. He side-stepped the debate concerning photons, and suggested that if electromagnetic radiation can behave as particles, perhaps particles behave like waves. He even worked out a way to calculate the wavelength that might be associated with a particle: it would depend on the particle's momentum, just as the wavelength of a photon depends upon a photon's momentum. De Broglie used the formula derived by Planck, and developed by Einstein, to work out what the wavelength would be for each electron in Bohr's allowed atomic orbits. He found that, in each case, the associated wavelength would fit an exact number of times around the circumference of the orbit. This was too much of a coincidence to be mere coincidence. It seems that the reason there are allowed orbits for electrons in atoms is that electrons behave like waves.

In 1927, experimental evidence was obtained that supported de Broglie's idea. American physicists Clint Davisson (1881 - 1958) and Lester Germer (1896 - 1971) fired electrons at a crystal of nickel metal; the electrons reflected from atoms in the crystal, and were collected by a detector. This procedure had been carried out with X-rays – very short wavelength electromagnetic radiation. The X-rays bounced off the atoms in the crystal, and made an impression on a photographic film. The result with X-rays is an interference pattern. In other words, many X-rays are detected at some angles, while far fewer were detected at others. The positions of the 'dark' and 'light' areas depend upon the spacing of the atoms in the crystal and on the wavelength of the X-rays. In terms of a wave interpretation of this result, the dark areas correspond to positions at which the waves 'cancel out'. In exactly the same way, light waves cancel each other out in forming interference patterns such as those that give rise to the coloured patterns on oil films or the surface of a soap bubble. Since X-rays are a form of electromagnetic radiation, this came as little surprise.

When Davisson and Germer bombarded a crystal with electrons, rather than X-rays, they found the same kind of interference pattern, with the spacing between the fuzzy 'dark' and 'light' fringes depending exactly on the ‘wavelength’ that de Broglie had suggested for the electrons. So de Broglie’s strange idea was correct. Electrons, which had always behaved like tiny particles, were here also behaving just like waves. By firing the electrons at different speeds, Davisson and Germer could vary the momentum, and therefore change the wavelength, of the electrons; the interference pattern changed accordingly. Other experiments, similar to Davisson and Germer’s, extended the idea of ‘matter waves’ to entire atoms. In 1999, another experiment demonstrated the phenomenon, but with large molecules, each made of 60 atoms. So, even composite objects like atoms molecules have a wavelength associated with them. In fact, every object has an associated wavelength. But for objects on everyday scales, that wavelength is tiny: far smaller than the size of an atomic nucleus. This is why we are not familiar with the strange effects observed with microscopic particles such as electrons and atoms.

During the 1920s, physicists formulated mathematical methods which addressed the wavelike nature of matter, and which aimed to predict the behaviour of any small-scale physical system. It was an attempt to rework ‘classical’ physics so that it would fit into the newly-discovered quantum nature of matter and energy. The result was to be Quantum Mechanics. Three very different mathematical formulations of Quantum Mechanics were developed, which were all later shown to be equivalent. One of them, called wave mechanics, was put forward by Austrian physicist Erwin Schrödinger (1887 - 1961) in 1926. Since matter behaved like waves, Schrödinger developed a wave equation for matter. The main feature of Schrödinger's approach is the wavefunction: a mathematical description of everything that is known about a system. The wavefunction describing a particle bears a strong resemblance to an electromagnetic wave. Just as the strength, or amplitude, of an electromagnetic wave determines how intense the wave will be at a certain point and a certain time, the wavefunction determines how ‘intense’ the particle will be at a certain point and time. In 1926, German physicist Max

Born (1882 - 1970) realised that this interpretation carries with it something profound. The wavefunction does not describe the exact position or momentum of a particle: rather, it describes the probability of a particle being in a certain position and the probability of the particle having a certain momentum.

To understand how the wavefunction is related to probability, consider a single photon emitted by an atom. In Maxwell's theory, this would be interpreted as an electromagnetic disturbance which spreads out in all directions at the speed of light: a wave that becomes weaker the further it travels from the atom. You can think of this electromagnetic wave as being the wavefunction of the photon. The wave has the same intensity in all directions; this is equivalent to saying that the probability of finding a photon in any direction around the atom is the same. What would happen if you set up a number of identical detectors surrounding the atom? Maxwell's theory would predict that each detector would register a tiny proportion of the total electromagnetic disturbance. According to Quantum Mechanics, however, only one detector will respond, and that detector will register the entire energy of the photon. Since the wave is travelling out equally in all directions, each detector has the same probability of detecting the photon.

An uncertain Universe A wavefunction describes all the possible states of a physical system, and assigns a probability to each one. Until an experiment is carried out or an observation is made, the wavefunction exists as a mixture of all the possible states, the most likely states making the greatest contribution. Mathematicians call this kind of mixture of waves a superposition. Consider the electrons forming an interference pattern in Davisson and Germer's experiment. The interference pattern produced is like a map of the probability of each electron's final resting place. Indeed, if the experiment is carried out one electron at a time, each electron arrives at a specific point, and the pattern builds up gradually. In a similar way, only by throwing a pair of dice a large number of times can you demonstrate the different probabilities associated with each possible score. The score for a particular throw is random and unpredictable, and is determined solely by probability. In Davisson and Germer's experiment, the 'darker' areas of the pattern correspond to low probabilities, the 'lighter' areas to high probabilities, and the fuzzy fringes in between correspond to an intermediate probability. According to the standard interpretation of the wavefunction, the electron itself exists in all possible trajectories until it is detected. At the moment of detection, the wavefunction 'collapses' to just one of the possible outcomes.

Despite the radical nature of his own theories, Einstein was uncomfortable about the emphasis on probability in Quantum Mechanics. Classical physics was built on the cosy assumption that particles behave strictly according to well-defined laws of physics. In this 'deterministic' picture, there was no room for concepts such as the wavefunction, superposition and randomness. The emerging discipline of quantum physics was insisting that the only deterministic feature of a particular system of particles is the ratio of probabilities of all possible outcomes. The actual outcome of a particular experiment on the system at the microscopic level is as random as the roll of a dice – and only becomes well-defined over a large number of 'throws'. In practical terms, this is not too troubling, since physicists can use the mathematics of quantum mechanics to produce extremely accurate predictions of the behaviour of systems of particles. But in philosophical terms, quantum mechanics is extremely troubling. What is it that controls exactly what a particle does? Where are the Universe's dice? Many different interpretations of quantum physics have been put forward to explain what might be going on, every one somehow philosophically troubling.

One of the most popular interpretations of the strange world of the quantum (the most useful and mathematically rigorous) is the 'sum over histories' approach, developed by American physicist Richard Feynman (1918 - 1988). In this approach, you work out the probability of a certain event by considering every possible way it could have happened. Consider, for example, Davisson and Germer's experiment, in which an electron bounces off an atom in a crystal and lands on a screen. A particular outcome of this experiment, the electron hitting a particular point on the screen, can have a number of different 'histories'. Each one is a sequence of events that ends with the electron arriving at the required position at the required time. The most simple history involves the electron travelling in a straight line, being repelled by the electrons in the crystal's atoms, and bouncing back up to the screen. But it is also possible that the electron bounces off several different atoms before reaching the screen. It is also possible that the electron travels in trajectories that are not straight; in fact, almost anything is possible. Each history is made of a number of stages, each one a fundamental event. To calculate the probability for a particular history, the probability of each stage is taken into account. To work out the overall probability that the electron will hit a certain point on the screen, all the possible histories are considered, and the probability of each is added to make a sum, or 'path integral'. In some places on the screen, the value of the integral will be zero, in others, it will be nearly one, and in others, it will be somewhere between the two. This process is laborious, and can only be applied practically to very simple cases. Moreover, only the probabilities of the simplest histories can be calculated. Fortunately, the simplest histories provide the largest contributions to the overall probability. But Feynman's scheme does little to quell philosophical distress with the quantum world: Does the particle really travel through all of its histories, all possible paths? Perhaps it has a way of looking at each path and deciding which one it wants to go along. Or perhaps every particle carries with it a pair of dice.

In fact, wave-particle duality suggests another fact, equally challenging to the notion of determinism: it is impossible to determine both the momentum and position of a particle with absolute accuracy. The degree of uncertainty in a particle's momentum multiplied by the degree of uncertainty in its position is always greater than a certain number. According to classical physics, any uncertainties in the measurement of a particle's position and momentum are due to inaccuracies in the equipment making the measurement. The limit on measurement was realised in 1927, by German physicist Werner Heisenberg (1901 - 1976). The Uncertainty Principle, as this idea is known, arose from one of the alternative formulations of Quantum Mechanics, called matrix mechanics. Heisenberg's Uncertainty Principle sounds abstract, perhaps even unimportant. Why would we ever want to know the position and momentum of a particle with complete accuracy anyway? The real significance is not a practical, but a philosophical one. Heisenberg's finding is fundamental to our understanding of the Universe, and is a central feature of the weird world of quantum physics. And, as we shall see, it explains the production of the virtual particles that are central to Stephen Hawking's explanation of black hole radiation.

The real meaning of Heisenberg's principle is hidden in the formulation of 'Matrix Mechanics' – although it was quickly realised that it lurks in the other two formulations of Quantum Mechanics, too. Heisenberg initially discovered it as a mathematical relationship, and originally called it the 'uncertainty relation'. He discovered, too, that there are other pairs of quantities whose uncertainties are intrinsically linked. Energy and time are the most important, and feature prominently in the modern physicist's understanding of the Universe, as we shall see. Heisenberg's discovery raises important questions about what it means for something to exist, since measurement of something gives meaning to it. The uncertainty relation can also be derived by considering the effect that measurement has on a system. Roadside speed cameras measure the speed of a car by bouncing radio waves off it. The momentum of the radio waves have no measurable effect on the speed of the car. But at the subatomic level, the momentum of photons do affect particles, and must be taken into account. Any measurement involves an interaction between particles. This fact blurs the line between the observer and the observed: in the quantum world, everything has a tangible effect on everything else. However one wishes to interpret the Uncertainty Principle, it remains a fact of nature, and it has some strange but enlightening consequences.

One illustration of Heisenberg's Uncertainty Principle is given by an electron confined in a long, very narrow box. Beyond the ends of the box, the wavefunction of the electron must be zero, since there is zero probability of finding the electron there. This also means that the value of the wavefunction at each end of the box must be zero. There is a large number of waves, each with a different wavelength, that fit this criterion. Since only certain waves are allowed, the electron can only possess certain energies. The lowest allowed energy is determined by the longest allowed wavelength. This turns out to be a wave whose wavelength is twice the length of the box – since a wave has a zero point halfway along its length. The important point is that the minimum energy is not zero. And so, even if the box and the electron are cooled to the theoretical minimum temperature, absolute zero, it will still be moving to and fro in the box. The 'zero point energy' in this case depends upon the length of the box. The wavefunction does not specify in which direction the electron moves, so there is an uncertainty in the momentum of the electron. The zero point energy, and therefore the uncertainty in momentum, is highest when the box is shortest. And when the box is short, the uncertainty in position is the least. If, on the other hand, you make the box infinitely long, so that you have no idea where the electron is, the lowest energy corresponds to an infinite wavelength. In this case, the momentum is well-defined: it is zero.

New fields Heisenberg's Uncertainty Principle is a consequence of wave-particle duality: the fact that matter and electromagnetic radiation can behave both as particles and waves. Two main formulations of quantum mechanics approached this duality in different ways. Wave Mechanics was built firmly upon the wave picture, while Matrix Mechanics was devised more with particles in mind. In 1927, Niels Bohr realised that both interpretations must be correct, but could be applied in different circumstances. In some experiments, an electron behaves like a particle, in others it behaves like a wave; the same is true of the photon. Wave-particle duality is uncomfortable, but it is real. We have an idea of what a wave is, in the context of fields, but what is the real nature of a particle? What is matter really made of?

The atomic theory of the 19th century was based on the idea that atoms are the smallest constituent of matter. It was assumed that they would be solid and impenetrable. The name 'atom' reflects this: it derives from the Ancient Greek word *atomein*, meaning ‘uncuttable’. However, atoms are 'cuttable', and they are far from being the smallest constituent of matter. Throughout the twentieth century, physicists have uncovered a plethora of previously unknown particles smaller than the atom, some of which have been produced artificially in particle accelerators. The first sub-atomic particle to be identified was the electron, in 1897. The reason why the electron was thought-of as a particle when it was discovered is that it has well-defined values for its mass and its electric charge. The discovery of radioactivity in 1900 suggested the existence of other tiny particles, again with definite masses and electric charges, which seemed to originate from within the atom. In 1911, Ernest Rutherford discovered that the centre of the atom is small, dense and positively charged. Atoms of different elements have different amounts of positive charge in their nuclei, each one a multiple of a fundamental unit of charge. The nucleus of a hydrogen atom has a charge exactly equal to that fundamental charge. So, in 1914, Rutherford suggested that all nuclei were made of a number of individual, positively charged particles, each with the same charge as the hydrogen nucleus. He called the particles protons, and surmised that each element is defined by the number of protons in its nucleus. He went on to predict the existence of a particle that consists of a proton and an electron combined, which he called the neutron. Rutherford's motivation in this was to explain why protons stay clumped together in a nucleus: protons are all positively charged, so they repel each other. Rutherford proposed that neutrons, with no net electric charge, might somehow help to prevent the nucleus from flying apart. He was right, although when the neutron was discovered in 1932, it was hailed as a particle in its own right, rather than a combination of a proton and an electron.

Despite the unresolved issue of what a particle really is, physicists were convinced that the electron, the proton and the neutron were the real 'atoms' of matter. They believed that these particles are truly uncuttable: the fundamental ingredients of all matter. However, quantum theory was about to rock that view, through 'relativistic quantum physics'. The fact that subatomic particles can travel very fast in some circumstances meant that physicists would have to incorporate Special Relativity into Quantum Mechanics. The resulting theory is called 'Quantum Field Theory'. According to Quantum Field Theory, the wavefunction itself is thought of as a field, similar to the electromagnetic field. A particle can be seen as a quantum of that field – just as a photon is a quantum of the electromagnetic field. So, there must be an electron field, a proton field and a neutron field, with each particle existing as a quantum of its respective field. This is similar to the idea Faraday had had nearly one hundred years earlier. If it is true, there is no such thing as a solid, impenetrable particle – clearly a radical departure from the mechanistic view of nineteenth century physics. Instead, all matter is made of interacting fields, and what we think of as particles are just individual quanta of energy within them.

In 1927, British physicist Paul Dirac (1902 - 1984) constructed an equation that described an 'electron field'. The equation predicted the behaviours of electrons very well, but it had another, unexpected result. It seemed to predict the existence of electrons with negative energy as well as electrons with positive energy. Dirac got around this apparently nonsensical result by suggesting that all of the negative energy states were filled, a kind of constant sea of electrons. His odd new negative energy electron could then be interpreted as a hole or bubble in the electron sea. In the real world, it should appear as a positively charged particle with the same mass as the electron. And, in 1932, Dirac's particle was detected. This 'anti-electron' was named the positron, and since 1932, countless positrons have been observed by the tracks they leave in cloud chambers or on photographic film. In all cases, they have been produced as a result of the decay of other particles; positrons do not fit into the hierarchy of ordinary matter, unlike the electron. It was later shown that every type of particle must have a corresponding anti-particle, although in some cases, the particle can be considered as its own anti-particle; an example is the photon.

Dirac approached his electron field equation with the condition that it must incorporate a property of electrons called spin. An electron behaves as if it is spinning, like a top – an observation that had been made by physicists studying the finer details of electron energy levels in atoms. Like other particle properties in the quantum world, spin is quantised. A particle can either have integer spin (-2, -1, 0, 1, or 2 times the unit of spin) or half-integer spin ($^1/_2$, -$^1/_2$ etc). An electron has $^1/_2$ or -$^1/_2$ times the unit of spin. In his equation, Dirac found that particles with half-integer spin are 'exclusive': there can only be a single electron in each allowed state. This made sense in the context of atomic energy levels: if more than one electron could occupy the same energy level in an atom, then all the electrons would naturally tend to be in the lowest energy state. The fact that they are not was first noted by Austrian physicist Wolfgang Pauli (1900 - 1958), who proposed his Exclusion Principle in 1924. Other examples of particles with half-integer spin are the neutron and the proton. The behaviour of these exclusive particles with half-integer spin is described by a mathematical scheme called Fermi-Dirac statistics, since it was worked out by Dirac and Italian physicist Enrico Fermi (1901 - 1954). Particles that obey Fermi-Dirac statistics are called fermions. The exclusivity of electrons accounts for the electron degeneracy that supports white dwarf stars against gravitational collapse, as described in Chapter One. In fact, all particles with the same amount of spin as electrons behaves exclusively in this way. An example is the neutron – and the degeneracy of neutrons is what holds up neutron stars against collapse.

On the other hand, particles with integer spin, such as the photon, can be described by an equation similar to Dirac's, called the Klein-Gordon equation, after Oskar Klein (1894 - 1977) and Walter Gordon (1893 - 1940). The Klein-Gordon equation was originally designed to describe the behaviour of electrons, but it was not adequate for that task. Particles with integer spin, which are described by the Klein-Gordon equation, obey Bose-Einstein statistics, developed by Einstein together with Indian physicist Satyendranath Bose (1894 - 1974). They are called bosons, and they are not exclusive like their distant cousins, the fermions. Under the right conditions, the wavefunctions of many bosons can merge together, to form a composite wavefunction that describes a single particle. Systems made up of many particles, such as atoms, will be bosons if the spins of their component particles add up to an integer. In some atoms, the spins of all the fermions (electrons, protons and neutrons) add up to an integer value, and the atom as a whole behaves as a boson. If they are made cold enough, many such bosonic atoms can 'condense' into a single-wavefunctioned amalgam called a Bose-Einstein condensate. This strange state of matter, one single particle consisting of many individual atoms merging together, was predicted in the late 1920s, and first created in the laboratory in 1995. It is worth pointing out that if all matter were made of bosons, rather than fermions, there would be no such thing as solids. It is because of the exclusion principle that we do not fall through the floor.

So, during the 1930s, it seemed that matter is made of fermions: electrons, protons, neutrons and positrons. Photons, which are bosons, clearly played an important role, and could be considered as much a particle as any of the others. Quantum Field Theory, which brought about this new division between fermions and bosons, and predicted the existence of the positron, also had something very important to say about the role of photons.

Virtual reality When photons are absorbed by electrons in atoms, they cease to exist. And when they are emitted by electrons, they are created. These annihilations and creations are incorporated into the mathematical formulation of Quantum Field Theory. According to Heisenberg's Uncertainty Principle, there is a direct relationship between uncertainty in energy and uncertainty in time, just as there is between the uncertainty in position and the uncertainty in momentum. This means that particles can be created from nothing, borrowing the energy necessary for their existence as long as they are annihilated within the corresponding time. So, pairs of electrons and positrons can be created and annihilated throughout spacetimespacetime, through an unobservable virtual oscillation of the electron field. And, because this behaviour is possible, it makes a contribution to the wavefunction of any system. The Uncertainty Principle therefore demands that Dirac's electron sea is a choppy one, both above and below the surface. This explains the 'virtual particle pairs' that Stephen Hawking uses to explain how a black hole can radiate. The idea of virtual particles is a consequence of Quantum Field Theory, so all fields have virtual fluctuations. Empty space, crossed by fields, must be a seething foam of virtual particles, each only existing for as long as the Uncertainty Principle allows, and not long enough to be detected. The concept of virtual particles is alien, perhaps living only in the mathematics that describes reality, rather than in reality itself. But the importance of virtual particles goes far beyond an explanation of black hole radiation and electron-positron pairs. Virtual particles can explain the mechanism behind electromagnetic forces.

It was known that electromagnetic forces are transmitted by the electromagnetic field; that was how Faraday had defined the concept of a field in the 1840s. It is common to think of electromagnetic forces as only being important when you experiment with static electricity, with magnets and electromagnets, or perhaps when you use an appliance that has an electric motor in it. But the force of electromagnetism is far more important and ubiquitous than that. The atoms that make up the world around you are held together by the electromagnetic forces between the electrons of neighbouring atoms – so without electromagnetism, nothing would be solid. In fact, atoms themselves are held together by the electromagnetic force between negatively-charged electrons and positively-charged protons in the nucleus. So, how do virtual particles help to explain all of this?

To explain the force between, say, two electrons, Quantum Field Theory supposes that the electrons continuously exchange virtual photons. The photons carry momentum, which pushes the electrons apart. This is why we observe pairs of like charges repelling each other. The same mechanism can be used to explain why unlike charges attract, like the balloon and the jumper described earlier. Electromagnetic forces like these are stronger the closer the charged particles are. This can be explained by the fact that it takes less time for the virtual photons to travel between particles that are closer together, so they can have higher energies, according to the Uncertainty Principle. The range of the electromagnetic force is, in principle, infinite. Since the virtual photons involved in electromagnetism are massless, they can borrow a tiny amount of energy for a very long time. This enables them to pass between charged particles that are separated by huge distances, albeit supplying a tiny amount of momentum. (In ordinary matter, the electric charges balance out overall, which is why only gravity dominates at large scales.) The complete Relativistic Quantum Field Theory of electromagnetic interactions is called Quantum Electrodynamics. It is physics' most accurately tested and reliable scientific theory to date. Might other interactions between the particles of matter be effected in this way, by virtual 'exchange particles' in quantised force fields?

At the time that Quantum Field Theory was being developed, physicists knew of only two other interactions between the particles of matter besides electromagnetism. Gravity, of course, is one of them. General Relativity demonstrated that gravity is not a force in the same way as electromagnetism is: it is a result of the warping of spacetime. Nonetheless, gravity is still an interaction, and General Relativity is a field theory. It seems reasonable to suppose that it is quantised – although to this day, that remains only a supposition. The hypothetical quantum of the gravitational field is called the graviton – which, since gravity has an infinite range, should be massless, like the photon. The other force known at the beginning of the 1930s was called the nuclear force, which is only attractive, not repulsive. Its existence was postulated because something had to be holding protons together against their electromagnetic 'will' in the nucleus. Rutherford had proposed that neutrons must have something to do with this force, and he was right. In fact, both protons and neutrons are affected by the nuclear force, and it is this force which holds the nucleus together. Over distances comparable to the diameter of a nucleus – 0.000000000000001 m (one thousand-trillionth, 10^{-15}, of a metre) – the nuclear force is clearly very strong. But outside the nucleus, it drops practically to zero: if it did not, then nuclei would be larger than they are. The very limited range of the nuclear force suggested that the virtual particle that carries it has mass, unlike the photon and, if it exists, the graviton. To see why, remember that a particle with mass has energy even if it is stationary, by virtue of the mass-energy equivalence. This puts a lower limit on the energy that can be borrowed by the Uncertainty Principle. In turn, this puts an upper limit on the time the virtual particle can exist, and therefore the distance across which the virtual particles can travel. So, simply by knowing the range of the nuclear force, it should be possible to estimate out what mass the exchange particle should have. In 1934, Japanese physicist Hideki Yukawa (1907 - 1981) did just that for the virtual carrier particle of the nuclear force. In 1947, a particle was observed that met all the requirements of Yukawa's prediction. It is called the pi meson, or 'pion'.

Although virtual pions are indeed exchanged between protons and neutrons, and do keep the nucleus together as Yukawa supposed, the pion was later found to be made of smaller components, called quarks. The nuclear force that pions carry turned out to be a 'residual' effect of a more fundamental interaction: the strong interaction. All quarks experience the strong interaction, and it is therefore essential in explaining how particles made of quarks behave. And not only pions, but also protons and neutrons, are made of quarks. The existence of quarks, predicted in the 1960s by Murray Gell-Mann (born 1929), is therefore a central feature of modern physicists' understanding of the hierarchy of matter. Any particles that are composed of quarks and interact via the strong interaction are called hadrons. Protons, neutrons and pions are examples of hadrons, but there are many more possibilities. Electrons, however, are not hadrons: they are completely unaffected by the strong interaction. Any particles that are unaffected by the strong force, including electrons, are categorised as leptons.

So, the slightly more complicated picture of the atom now consists of a nucleus made of two types of fermion (protons and neutrons) that are also hadrons, surrounded by other fermions (electrons) that are leptons. The atom is held together by the exchange of virtual photons (which are bosons) between the electrons and the protons; the nucleus is held together by the exchange of virtual pions (another type of boson) among the protons and neutrons. The protons and neutrons – and indeed, the pions – are themselves composed of smaller particles still: quarks. So, what is a quark, and how do quarks join together to make protons and neutrons?

Yet more particles There are six different types, or flavours, of quark, and six corresponding anti-quarks. To make the categorisation of the particles of matter even more bewildering, the hadrons are further divided into two classes: mesons and baryons. Mesons, such as the pion, are each made of two quarks, while baryons, such as protons and neutrons, are composed of three quarks each. Surprisingly, the quarks that make up the proton and neutron are not tightly bound together. They can move around freely, since unlike the electromagnetic force, the strong force actually increases with distance. At near-zero separation, the force is negligible, but it increases extremely rapidly over its tiny range (and then drops off almost as rapidly). So, while the quarks are not tightly bound, they are very strictly 'confined'. Quarks are so strictly confined that they are never observed individually: they are always in their pairs (mesons) or their triplets (baryons). Just as the electromagnetic interaction is carried by virtual photons, the force between quarks – the strong interaction – is carried by particles called virtual gluons. These are bosons, just like photons (and gravitons, if they exist). Both quarks and gluons carry something called 'colour charge'. While two types of electric charge (positive and negative) are required to explain the electromagnetic interaction, three types of 'colour charge' are needed to explain the strong interaction. Well-defined rules govern the interactions between coloured quarks, anti-quarks and gluons. In fact, since gluons carry colour charge, they can interact with each other. It is thought that this leads them to form into bundles called glueballs, although no one has yet observed one of these odd objects.

The strong force becomes extremely strong with increasing separation between quarks, but what if you pull the quarks apart hard enough to overcome the force – try to tear the proton into its constituent parts, for example? It is possible to separate quarks, but a strange thing happens when you do. Imagine pulling apart the two quarks of a meson: they will resist, but if you supply enough energy, a new quark-antiquark pair will be created from the energy you supply. Each member of the new pair moves off with one of the original pair, forming new, separated mesons. This is a little like pulling a rubber band until it breaks in the middle: you are left with two rubber bands, both with two ends, and the elastic forces are once again contained within the two new relaxed pieces of rubber. In the case of the rubber band, however, the total mass of the two new pieces formed is the same as the original single piece. This is not the case for the torn meson: each new fragment is a complete meson in its own right. The extra mass comes from the energy put into the system to pull apart the meson.

The 'colour charge' theory of quarks and gluons – the relativistic quantum field theory of the strong interaction – is extremely successful in predicting the behaviour of hadrons (particles affected by the strong interaction). It is called Quantum Chromodynamics, because it is very similar to Quantum Electrodynamics, the quantum field theory that describes electromagnetism. Quarks are also involved in the fourth and final known interaction between particles in our Universe: the weak interaction. Unlike the strong interaction, which can be thought of as a force, the weak interaction merely involves the change of flavour of fundamental particles. For example, the weak interaction explains one kind of radioactive decay, beta decay, in which a neutron becomes a proton. A neutron would be a proton if just one of its quarks changed its flavour. Under the influence of the weak interaction, this is exactly what happens. In the process, an electron is created – and yet another different particle, the neutrino, is produced, too. The neutrino, like the electron, is classed as a lepton, since it is unaffected by the strong interaction. I shall discuss this elusive particle in chapter five. Any example of the weak interaction involves either hadrons (the particles containing quarks) or leptons (particles such as the electron and the neutrino), or both. In fact, the weak interaction is the only one, except gravity, that affects all known classes of particle.

Like the strong interaction and the electromagnetic interaction (and probably the gravitational interaction) the weak interaction is mediated by virtual particles. There are three different particles: the W^+, the W^- and the Z, and once again, these are bosons. These exchange particles are very massive: they have about the same mass as an entire medium-sized atom containing tens of protons and neutrons. According to Heisenberg's Uncertainty Principle, as explained above in the case of the pion, this makes the range of the weak interaction extremely short. In fact, its range is about one thousandth of the width of a proton: about one million-trillionth of a metre (10^{-18}m).

There is another boson that you may well have heard of, in addition to the photon, the gluon, the two Ws and the Z and (possibly) the graviton: the Higgs boson, named after British physicist Peter Higgs (born 1929). As with the others, the Higgs boson is associated with a field – the Higgs field – and it is interaction with that field that gives other fundamental particles their mass. When it was discovered, in 2013, the Higgs boson provided important evidence in favour of the modern picture of fundamental particles and their interactions: the 'Standard Model'. According to the Standard Model, there are two classes of particle that can be called fundamental: the true 'atoms' of our Universe. They are quarks – which are the constituents of hadrons such as the proton – and leptons, such as the electron and neutrino. All of the exchange particles that facilitate the interactions are bosons, and are more properly referred-to as gauge bosons. The Standard Model makes good sense out of the apparent chaos, arranging the known particles neatly into families, each with three distinct generations. However, physicists would like to make even more sense of it, by unifying the matter and forces under a single theory. A 'Theory of Everything' would ideally describe all particles as aspects of a single field. The quest to find a single, Unified Field Theory, is the ultimate concern of modern physics. And so, in the next chapter, we shall look at the most promising avenues that physicists are exploring in this quest.

5: TYING IT ALL TOGETHER

The Standard Model of the fundamental particles of matter (fermions) and their interactions (via virtual bosons), outlined in Chapter Four, has proved hugely successful in explaining the hierarchy and the behaviour of matter in our Universe. The main reason for its triumph is that it addresses the requirements of both Special Relativity and Quantum Mechanics – two very excellent theories in their own right – but goes beyond both of them. Through sophisticated mathematics, it gives very specific and accurate descriptions of particle behaviours in almost any given situation. The Standard Model has even enabled theoretical physicists to predict the existence of previously unknown particles, and has given experimental particle physicists the knowledge of how to find them, the Higgs boson being the latest example. Neat and successful though the Standard Model is, it is not a complete description of reality: there are several fundamental physical constants that it cannot predict, and which have to be input 'by hand' from experimental results, rather than emerging directly from the theory. For example, the Standard Model provides no way to calculate the values of electric charge carried by a proton or an electron, nor does it explain why they should be the same. In this chapter, we shall begin by investigating the successes and pitfalls of the Standard Model.

The other main theory that physicists have for describing reality is General Relativity, described in Chapter Three, which relates the distribution of mass-energy to the geometry of spacetime. That theory is also extremely successful – but it, too, has its pitfalls. It is likely that any theory that can integrate the quantum mechanical nature of the Standard Model with the continuous spacetime geometry of General Relativity will be a 'Theory of Everything'.

Crash scene investigations How can physicists test a theory that describes the behaviour of particles of matter that are far too small to be seen? Fortunately, the Standard Model can be used to compute very specific predictions about the outcomes of particle interactions, and these predictions can be evaluated by conducting experiments in particle accelerators. Inside a particle accelerator, electrically-charged particles are made to travel at incredibly high relative speeds, propelled, guided and focussed inside evacuated metal tubes by intense electric and magnetic fields. Accelerators with straight tubes are called linear accelerators. There are circular accelerators, too, in which the accelerating fields are provided by bursts of microwave radiation that boost particle speeds at several points around a ring. The particles travel in bunches, surfing the crests of the microwaves. Just like pushing someone on a swing, the boosts must be given at exactly the right time to have the desired effect, so circular accelerators are called 'synchrotrons'. In most cases, particles from a linear accelerator are injected into a synchrotron.

Inside the tubes of a synchrotron, beams of oppositely-charged particles travel in opposite directions, and can be made to cross each other's path. In a linear accelerator, the beam of particles typically strikes a stationary target. In each case, the result is a large number of high-speed, high-energy collisions. The portion of the accelerator tube where the collisions take place is called the collision chamber. If the particles involved in collisions are made up of smaller parts – like the proton, which is made of quarks – then collisions can bring to light evidence of their internal structure. If the collisions have enough energy, they can even break the particles into their constituent parts. In the case of a proton, this means separating quarks from one another's grip. As we saw in

Chapter Four, when you separate quarks, the energy input transforms into a new quark-antiquark pair. This explains why, to date, no single quarks have been observed, while the debris from such an interaction always includes mesons (quark-antiquark pairs).

Mesons produced in a collision decay after a tiny fraction of a second, as a result of the electromagnetic or weak interactions, typically producing either muon-antimuon pairs or gamma rays, accompanied by neutrinos. In turn, the muons and the gamma rays may themselves decay, into electrons and positrons. Each step involved in this kind of event can be understood in its finest details in terms of the Standard Model, and therefore collisions act as a good test for the theory. The experimenters can deduce the identities of the particles present after the collisions from their behaviour in the strong magnetic and electric fields in the detectors that surround the collision chamber. Particles leave 'tracks' in detectors, and, like crash scene investigators, particle physicists examine those tracks. Only charged particles bend in the electric and magnetic fields, and only charged particles leave tracks. But there are two ways in which the existence of uncharged particles can be inferred. Firstly, many collision products – including uncharged ones – decay in a tiny fraction of a second, producing particle-antiparticle pairs. Typically, one member of such a pair carries positive electric charge, the other carries negative, so physicists can detect these collision products. Secondly, the total energy and momentum of the particles that do leave tracks can be calculated, and any energy unaccounted for should be predicted by the Standard Model as uncharged particles.

An example of an uncharged particle that is frequently produced in large numbers by collisions is the neutrino. There are three different types of neutrino, one for each generation of the particle family. They are the electron neutrino, the muon neutrino and the tau neutrino. The existence of the neutrinos was first postulated as a result of physicists evaluating the energy and momentum involved in beta decay, in which a neutron decays to form a proton. This process also results in the emission of a fast electron (in this case called a beta particle for historical reasons). The energy of the fast electrons produced in beta decay was found to have a continuous range of values, which was puzzling. Also, the recoil of the nucleus was not in the direction directly opposed to the flight of the electron. The explanation of these observations was that a new particle, the neutrino (actually an anti-neutrino in that case!), is emitted, carrying away an arbitrary amount of energy and momentum. Since neutrinos are extremely difficult to detect, their presence, their energy and their momentum could be inferred to make sense of the corresponding values for the particles that could be detected. This may seem like simply a convenient explanation – but, as it happens, it is correct. Neutrinos can now be created to order in particle accelerators and nuclear research laboratories, and can be detected, too. The reason why neutrinos are so difficult to detect is that they interact extremely weakly with other particles. They do not carry electric charge or colour charge, so they do not interact by the electromagnetic or strong interactions. They only interact with other particles through the weak interaction, which has a very short range, and through the gravitational interaction (and only extremely weakly, since neutrinos have a vanishingly small mass). Because of their elusive character, neutrinos make up the least-known class of particles. The tau neutrino was finally detected as recently as 2000.

I have already mentioned the Higgs field, and the particle that can be produced by disturbing it, the Higgs boson. The proof of the existence of this field was a crucial final piece in the puzzle of particles and fields. With the Standard Model jigsaw complete, you might think that physicists' work is done. However, as we shall shortly see, there are many unresolved issues, and the puzzle of the fundamental particles of the Universe remains unsolved.

A beautiful theory The Standard Model relies heavily on the principle of symmetry. In everyday language, we consider something as symmetrical if it looks the same when viewed in a mirror. The letter 'M', for example, is symmetrical in this way, while the letter 'L' is not. But symmetry has a deeper meaning, one that is intimately connected with all the basic laws of the Universe. An object or a system is symmetrical if it remains the same after you perform some action, or operation, on it. So, play a melody made up of musical notes, then play it again with all of the notes shifted up by a semitone, and the melody sounds the same. Here, a melody can be said to be 'invariant' under a key change, and musical keys therefore possess a kind of symmetry. Similarly, the letter 'M' can be declared invariant under a reflection. In general, the laws of physics are invariant under translations (shifts) and rotations in space and time. The Universe has intimate symmetries that demand that its laws do not recognise any 'frame of reference' as different from any other. This does not mean that two observers will make exactly the same measurements in a particular experiment. Instead, it means that the same laws can be used to predict the outcomes of experiments wherever and whenever they are conducted.

In a sense, the history of physics has been a journey towards the realisation of symmetry in the Universe. As we saw in Chapter Two, the Ancient Greek philosopher Aristotle believed that the Earth was fixed at the centre of the Universe. Several hundred years later, Danish astronomer Nicolaus Copernicus realised that the Earth is not at the centre of the Universe, because it is in orbit around the Sun. Astronomers slowly began to realise that even the Sun is not in a particular place in the Universe. And, in the 1920s, American astronomer Edwin Hubble showed that even our galaxy is not in a special place. His discovery of the expansion of the Universe, which we shall investigate in more detail in Chapter Six, suggests that every location of space was once in the same place, and therefore that every location must be equivalent.

In 1915 German mathematician Emmy Noether (1882 - 1935) proved that each symmetry observed in physical systems in our Universe gives rise to a law of conservation. For example, the Law of Conservation of Energy, which states that the energy of a system is constant, turns out to be a direct result of the invariance of the laws of physics under movement through time. The Law of Conservation of Momentum is a result of the invariance of the laws of physics under translations through space. And the Law of Conservation of Angular Momentum is a consequence of invariance under rotation in space. These laws – of classical (pre-quantum) physics – depend upon 'continuous' symmetries. The laws of quantum physics, on the other hand, depend upon the Universe's 'discrete' symmetries. To understand the difference between continuous and discrete symmetry, consider a circle and a square. A circle has continuous rotational symmetry, since it remains the same if you rotate it by any angle. A square, on the other hand, has discrete rotational symmetry, since it is only invariant under rotations that are a multiple of 90 degrees.

Discrete symmetries buried in the equations of Quantum Field Theory imply the conservation of certain properties of microscopic systems. These include electric charge, quark colour charge, spin, 'baryon number' and 'lepton number'. This means that in a given event in a particle accelerator, only certain outcomes are possible: those that the Universe's symmetry bookkeeping system allows. For example, an electron and a positron have opposite lepton numbers of +1 and -1 respectively, a total of zero, and so there is no requirement that any leptons be present after they annihilate. After a collision of a proton (baryon) and an electron (lepton), however, the baryon numbers of all particles present must add up to 1, as must the lepton numbers. These conservation laws are built into the mathematical formulation of the Standard Model, through a sophisticated branch of mathematics called group theory. And as a result of the application of group theory to particle physics, the fundamental interactions emerge: they are the way in which the Universe retains its symmetry.

In order to explain the fundamental interactions (strong, weak, electromagnetic and gravitational) in the Standard Model, physicists distinguish between 'global' and 'local' symmetries. A system with global symmetry is one that stays the same if all its component parts are changed to the same degree. Hold a rubber band and move it or rotate it as one piece, and it remains the same. Move various parts of a rubber band by different amounts and it will stretch or twist, however. Stretching and twisting a rubber band produces forces that act to retain the rubber band's original configuration: they are the Universe's way of retaining the rubber band's symmetry under local transformations. In Quantum Field Theory, a particle is described by a wavefunction, as explained in Chapter Four. A particle's wavefunction has two important properties at each point in spacetime: its phase and its amplitude. Phase is a mathematical property of all waves that describes what point of a wave's periodic cycle has been reached – so that, for example, two waves whose crests and troughs are matched are said to be 'in phase'. The amplitude determines the probability of finding the particle at a given position. Changing the phase of the wavefunction by the same amount throughout spacetime makes no difference to the amplitude; the wavefunction has global symmetry. For particles that are affected by the electromagnetic interaction, this results in the law of conservation of electric charge. For particles affected by the strong interaction, the relevant law is conservation of colour charge; for the weak interaction, it is a quantity called 'weak isospin'.

While global symmetries are the source of certain conservation laws in Quantum Field Theory, local symmetry invokes the existence of (bosonic) force fields. Make a local change to the wavefunction's phase, and the equations predict that a force field must exist in order to protect local symmetry – in a similar way to the stretching of the rubber band. When the particle is, say, the electron, the force field predicted has exactly the same properties as the electromagnetic field. The mathematics even predicts the existence of photons. The local symmetries of the fields of the Standard Model are called gauge symmetries, and this is why the exchange particles in the fundamental interactions – particles such as the virtual photon – are called gauge bosons.

The application of symmetry principles to Quantum Field Theory is enormously successful. The exchange particles whose existence it predicts are all bosons, with spin 1, and behave in just the right way to account for the three fundamental interactions covered by the Standard Model. It even predicts the correct number of force carriers in each case: one for the electromagnetic interaction (the photon), three for the weak interaction (the two Ws and the Z) and eight types of gluon for the strong interaction. However, there is one major problem: according to the symmetry principle, all the gauge bosons should be massless. In the case of the electromagnetic and strong interactions, this prediction is borne out in reality, since the photon and the gluons are indeed massless. But for the weak interaction, the prediction falls down: the W and Z bosons have mass; in fact, each one has nearly 90 times the mass of the proton. Strange as it may seem, this negative result actually holds great promise for the Standard Model and the hope of going beyond it, towards a Theory of Everything.

Broken symmetries The masses of the W and Z bosons of the weak interaction are a result of 'broken symmetry'. There are several symmetries in the Universe that have been broken. For example, while there exists in principle a symmetry between particles and their anti-particles, the matter of our Universe seems to consist exclusively of particles. Physicists can produce anti-particles at will, albeit for a split second, in particle accelerators, and observe their fleeting existence as a result of collisions of high energy 'cosmic rays' received from space. But they are absent from ordinary matter – including, it appears, from stars in our own and in other galaxies. According to the Standard Model, particles and anti-particles were created in slightly unequal amounts in the early Universe: 1,000,000,001 or so particles for every 1,000,000,000 anti-particles. This asymmetry may be the result of a broken symmetry that physicists call CP violation. The letters 'C' and 'P' are mathematical operations carried out on particle wavefunctions to investigate certain symmetries. The 'C' stands for 'charge conjugation', and relates particles to their anti-particles; 'P' stands for parity, and relates left-handed and right-handed particles to each other. Here, handedness can be thought-of as mirror image symmetry, and for subatomic particles, it generally involves spin. There are many examples of left- or right-handed asymmetries at molecular level or above, such as the preferred right-hand twist in DNA molecules. However, at the microscopic scale of subatomic particles, the symmetry of handedness seemed to stand: in the first few decades of quantum theory, physicists found that parity was conserved in all the systems they studied. Parity conservation played a part in disallowing certain transitions between atomic energy levels, for example. The Universe seemed to make no distinctions between left and right at the subatomic level.

In 1956, that view changed – for the weak interaction, at least. Chinese physicists Chen Ning Yang (born 1922) and Tsung Dao Lee (born 1926) were trying to solve a problem involving a type of meson called the kaon. Yang and Lee reasoned that the way kaons decay involved a violation of parity – a broken symmetry – and they designed an experiment that set out to test their idea. The landmark experiment was carried out in 1957, by a team led by Chinese physicist Chien-Shiung Wu (1912 - 1997). It involved radioactive atomic nuclei aligned in a magnetic field, to ensure that they were all spinning the same way. When the nuclei decayed by the weak interaction, the beta particles (fast electrons) were all emitted in the same direction. If parity was conserved in this experiment, the fact that the spins of the nuclei were aligned would have made no difference to the direction in which the beta particles were emitted. So Yang and Lee were right: parity is violated in weak interactions. What they had actually discovered was something about neutrinos, which are always emitted as a result of beta decay. It turned out that neutrinos always have 'left-handed parity' while anti-neutrinos have 'right-handed parity'. Since these two particles are related via charge conjugation (each being the other's antiparticle), it seemed that the P violation (in beta decay) was always accompanied by a C violation (in the neutrinos/anti-neutrinos) – as if they made up for each other, in order to retain the overall CP symmetry. In this case, the combined operation 'CP' would be invariant. This does seem to be the case – nearly all of the time. Some physicists propose that it is because of the small violation of combined 'CP' symmetry in the weak interaction, perhaps about one in a million or so, that matter dominates over antimatter in our Universe.

The breaking of symmetry is very important to physicists for other reasons. In particular, if a symmetry is broken now, it may not have been in the past. In other words, the interactions we observe now may all be different facets of a unified, perfectly symmetrical scheme that existed early on in the history of the Universe. An example of symmetry-breaking in our everyday world is the change of liquid water into ice. Liquid water molecules have continuous rotational symmetry, and they can move in any direction unhindered. When water freezes, its molecules become fixed in regular crystal structures, which have only limited symmetries: the symmetry is broken. Ice is less symmetrical than water. Many physicists believe that something similar happened to the Universe. The expansion of the Universe suggests that long ago, its average temperature was very high, and some of its symmetries must have 'frozen out' as it has cooled. For example, physicists have shown that the weak and electromagnetic interactions are aspects of a single force, whose symmetry was intact in the early stages of the Universe. It was this unified 'electroweak' theory that predicted the existence and the masses of the W and Z particles that are the gauge bosons for the weak interaction.

The W^+, W^- and Z bosons are normally virtual particles. However, by supplying enough energy to the boson fields – an amount of energy equivalent to the mass of the bosons – physicists can make virtual particles real. In the same way, real photons – normally the virtual gauge bosons that carry the force between electrically-charged particles – are readily produced whenever electrons or other electrically-charged particles lose energy. We can detect the real photons as light or other electromagnetic radiation. You can think of a vibrating electron as having the effect of 'shaking photons loose' from the electromagnetic field. Particle accelerators are not necessary to make real photons, because photons are massless. But to make W and Z particles real, you need to shake their fields with much more energy. This is how these particles were discovered, in electron-positron collisions, in 1984. The scientists who developed the electroweak theory – American physicists Sheldon Glashow (born 1932), Steven Weinberg (born 1933) and Pakistani physicist Abdus Salam (1926 - 1996) – even predicted at what energy the symmetry between the two interactions is broken. Above that threshold, two separate Quantum Field Theories become one. By extension, it seems reasonable that the strong interaction, and even gravity, might be subsumed under a single theory that applies at even higher energies. And as particle accelerators become more powerful, there is hope that evidence may be gathered that could prove such a theory correct.

The question of why some fundamental particles such as the W and Z particles have mass, while others such as the photon do not, is perplexing. Imagine a proton in motion (relative to your frame of reference). As it is moving, it will have kinetic energy, and that will give it mass (mass-energy), according to Special Relativity. A proton that is stationary (relative to you) will still have some mass: its rest mass. Most of this mass results from the kinetic energy of the quarks and gluons that make up the proton. These particles are zooming around, confined by the strong nuclear force. But even the quarks have intrinsic mass, even though they are (apparently) truly fundamental, not composed of yet smaller particles. To understand what that really means, particle physicists had to appeal once again to mathematics and symmetries. The mathematics of Quantum Field Theory is necessarily complicated, and the number of matter and force fields in the Standard Model is annoyingly large. Despite the fact that physicists' quest is for simplification – to explain the complexities of the Universe in terms of a single field – the best way to explain the intrinsic mass of fundamental particles is to invoke yet another quantum field. In 1964, several scientists, notably the British physicist Peter Higgs (born 1929), proposed that the W and Z particles might take-on what we call mass by interacting with a field that permeates all of spacetime. Through this interaction, these particles would feel a resistance to acceleration, and this resistance is what we normally call 'inertia', the property that we identify with mass. Mathematical analysis showed that

Higgs' field would interact only with the particles we know to have mass (like quarks), but not those known to be massless (like gluons). The existence of the Higgs field was confirmed in 2013 by producing and detecting its quantum, the Higgs boson – which, itself, has a large mass.

Grand unification In the last two decades of his life, Albert Einstein spent much of his time and effort in searching for a single theory that could describe all of physics: a unified theory. Many physicists share Einstein's devotion to this cause. The successful unification of the electromagnetic and weak interaction in the 1960s was seen as an important step in the right direction. It means that there only three fundamental interactions are needed to describe reality: the strong interaction, the electroweak interaction and gravity. A theory that includes the weak, strong and electromagnetic interactions is called a Grand Unified Theory; if it includes gravity, too, it is referred to as a Supergrand Unified Theory or simply a Theory of Everything. In the 1970s, Sheldon Glashow together with another American physicist Howard Georgi (born 1947) constructed a theoretical framework that attempted to unify the electroweak and the strong interactions – a Grand Unification. Their framework established the equivalence of leptons and quarks, and proposed twelve types of boson that could transform lepton to quark and vice versa. It was inspired by the symmetry between the generations of leptons and quarks: there are the same number of particles in each. In the Georgi-Glashow scheme, The twelve carrier particles would have very high masses, and are normally collectively referred to as 'X', 'Y' or 'leptoquarks'. An example of an interaction that is possible under this scheme is a positron emitting an X particle and turning into a down quark. Because X particles must decay to form leptons and quarks, it is proposed that they were present in the very early stages of the Universe, and that they subsequently decayed to form the particles of matter that we see today.

One of the consequences of the leptoquark theory, buried in the mathematics, is that the proton is unstable. However, the theory predicts that the half-life of a proton, the time it takes for half a sample of protons to decay, is at least ten million million million million million years. So, the theory cannot be ruled out on this prediction alone, since the best estimate for the age of the Universe is a tiny fraction of that. However, it does give physicists the opportunity to test the theory, since in a collection of ten million million million million million protons, one would decay on average every two years. Several experiments have already been carried out to detect the decays of protons. The most ambitious so far was the Irvine Michigan Brookhaven (IMB) experiment. It involved a tank containing 8,000 tonnes of water, deep underground, which was monitored for telltale signs of proton decay for ten years. No proton decays were observed (although important research on neutrinos was carried out), and this sets a lower limit on the half-life of the proton. Another consequence of Grand Unification theories like the one that predicts the leptoquark is that they demand the existence of massive particles called magnetic monopoles. Without these particles, there is no mechanism by which the electromagnetic interaction could have become what it is today. But magnetic monopoles, unlike X particles, would not decay, and there should therefore be huge numbers of them in the Universe. None has ever been detected, and this poses a serious problem for the Standard Model, and – as we shall see in Chapter Six – the application of the Standard Model to cosmology.

There are several important problems with trying to push the Standard Model beyond its design. One is that the leptoquark-X-particle proposal introduces yet more new bosons into the scheme, which seems to make things more complicated, not less so. However, the main problems lie in the Standard Model itself. While it is an excellent tool for particle physicists to use in calculating the outcomes of particle interactions, it is incomplete. For example, there are several nagging, unanswered questions, including: 'Why are there exactly three generations of particles?'; 'Why does the electroweak symmetry break down at the energy that it does?'; and 'Why is there violation of the CP symmetry?'. These questions may sound esoteric, but they are of fundamental importance to theoretical physicists trying to rescue an incomplete yet extremely useful theory. One of the overarching problems that makes the Standard Model incomplete is the fact that it does not include a description of gravity. This issue applies to quantum theory in general, since in the world of the very small, gravity has an insignificant effect. However, there are many important situations, such as the interior of black holes and the early Universe, in which gravity is strong even over small distances. In these cases, both General Relativity and Quantum Physics fail. Since the Universe is definitely quantised, the problems lie as much in General Relativity as they do in Quantum Mechanics. I shall shortly investigate the various approaches that attempt to bridge the gap between the lumpy quantum world and the smooth geometry of spacetime. These approaches are collectively labelled as theories of 'Quantum Gravity'.

Another serious problem in the Standard Model involves something called 'renormalisation'. Inherent in Quantum Theory is the assumption that particles are 'pointlike'. The Uncertainty Principle refutes the idea that particles really are infinitesimal points, since then their position would be truly localised. However, the Standard Model still treats particles as points in a mathematical sense, and this treatment leads to severe problems. For example, according to Quantum Electrodynamics (the quantum field theory of electromagnetism) the space around an electron is populated by a cloud of virtual photons. Some of these are the mechanism by which the electron interacts with other charged particles. But some of them will be emitted and then reabsorbed by the electron itself. The closer to the electron you look, the more energetic will be the virtual photons, because according to the Uncertainty Principle, the more energy a virtual photon has, the less time it has to exist. Very close to the electron, the energy of the photons becomes infinite. The theory predicts other infinities, including the charge and the mass of the electron. Renormalisation is a mathematical trick, which involves ignoring the infinities and substituting the measured values for the energy, mass and charge. With these quantities correctly entered into the calculations, the theory produces incredibly accurate answers to any question asked of it. However unsatisfactory this procedure seems, it does show that the theory of Quantum Electrodynamics must reflect the nature of reality in some fundamental way. Because the process of renormalisation fixes the theory, Quantum Electrodynamics is said to be 'renormalisable'. In fact, Quantum Chromodynamics (the quantum field theory of the strong interaction) and the unified electroweak interaction are both found to be renormalisable, too.

An important step in the process of renormalisation is to impose a 'cut-off': a distance within which the contributions of virtual photons are ignored. The cut-off distance is arbitrary. There is, however, an ultimate cut-off distance, beyond which physicists can have no knowledge of physical processes. It is called the Planck length. This is a tiny distance: by comparison, a proton is about one hundred million million million times larger. The Planck length has a great significance in quantum physics, and especially in the search for a quantum theory of gravity. Its value can be calculated from three fundamental constants that appear in General Relativity and quantum theory, but it has several other important connotations. For example, in order to measure the position of an object, you need to use light or other radiation whose wavelength is no bigger than the size of the object. So, for example, while you can locate an aeroplane using radar beams whose wavelength is a few centimetres, you would not be able to locate an atom in this way. If you wanted to locate an object with a diameter of one Planck length, you would need to use radiation with a wavelength no larger than the Planck length. As explained in Chapter Four, the wavelength of a photon of electromagnetic radiation determines the energy of the photon. The energy of a photon with a wavelength equal to the Planck length is equivalent to a mass, according to Einstein's mass-energy equivalence. This mass is called the Planck mass. It turns out that an object with a mass equal to the Planck mass will have a Schwarzschild radius (explained in Chapter Three) equal to the Planck length. The Planck mass is about equal to the mass of a flea, so the Planck length is the size of the event horizon of a black hole with a mass the same as a flea. If you have followed this tortuous line of reasoning, you will see that the Planck length and Planck mass are fundamentally related, and represent the frontier between quantum physics and General Relativity. Indeed, most physicists consider the 'Planck scale' defined by these quantities as central to any theory of Quantum Gravity.

Physicists go further than this in their interpretation of the Planck scale. Gravity is by far the weakest of all the fundamental interactions, but the Planck length is the distance at which gravity becomes as strong as the others. In fact, physics at the Planck scale is likely to be fully unified and hold the full symmetry of the laws of physics. One other important quantity that defines the Planck scale is the Planck time, defined as how long it takes light to travel one Planck length. The value of the Planck time is about 5 hundred-million-trillion-trillion-trillionth of a second (5×10^{-44} s). Physicists cannot measure distances smaller than the Planck length, and times less than the Planck time, because to do so, they would have to use energies that would create a black hole, from which no signal can escape. So, the Planck scale sets fundamental limits on measurements, and on the laws of physics. Indeed, as we shall see in Chapter Six, cosmologists using the Standard Model can make convincing, informed guesses about the history of the Universe all the way back to the point when its age was equal to the Planck time.

Strings and superstrings Renormalisation enables the electroweak and the strong interactions to retain their functionality down to distances near the Planck length. Unfortunately, however, the same mathematical trick cannot be applied to gravity: because it describes a perfectly smooth spacetime geometry, General Relativity is 'non-renormalisable'. So there is no way to incorporate gravity into the Standard Model as it stands. Even for the other three interactions, the fact remains that renormalisation is a mathematical sleight-of-hand, and is therefore a little unsatisfactory. There is another way around the problem of infinite values for quantities calculated in the Standard Model: demand that particles are not pointlike after all. This is the defining strategy of the most promising approach to the problem of Quantum Gravity, called String Theory.

As with all modern physics, the real power and beauty of String Theory is to be found in its detailed mathematical formulation. However, the general idea can be visualised in a familiar system: a guitar string. Any string held under tension has certain modes of vibration: particular frequencies at which it can vibrate. These frequencies are determined by the length and the weight of the string, and the tension under which it is held. So, a long, heavy, loosely held string vibrates at low frequencies and produces lower notes, while a short, light, tight string vibrates much more quickly and produces higher notes. The lowest frequency at which a particular string can vibrate is called the fundamental, but there are higher allowed frequencies, too. String Theory is a radical departure from the Standard Model, because it suggests that every particle of the Universe is an allowed vibration of tiny one-dimensional strings. If it is correct, it will truly unify all the fundamental interactions. The strings can have free ends or they can be in the form of closed loops. A plucked guitar string quickly loses energy to the surrounding air molecules and to the body of the guitar, and soon stops vibrating altogether. The strings of String Theory, however, do not lose energy: they are the fundamental units of matter, so there is nothing to which they can lose energy. The strings of String Theory have dimensions on the order of the Planck scale, so they are incredibly tiny.

String Theory was first conceived in the late 1960s, under the name 'dual resonance model', and was born in an attempt to address problems with the theory of the strong interaction. Only in 1970 did physicists realise that the objects described by the theory were vibrating strings. The theory was originally formulated to describe the particles of matter: fermions. However, it soon became apparent that the mathematics was describing the particles that carry the interactions (bosons) instead. The theorists could tell the difference because of the amount of spin the strings would have: bosons have integer units of spin. One of the most encouraging aspects of the new theory was that it included a massless boson with two units of spin – exactly the characteristics of the hypothetical graviton, as required in any theory of Quantum Gravity. With the aim of the theory as the description of fermions, however, the bosonic string theory was quickly adapted. The inclusion of fermionic strings was achieved by allowing the strings to split in two and rejoin, so that every possible spin state was included. One important consequence of this idea was the suggestion that there might be a host of new particles. Specifically, there would be a new type of boson corresponding to each known type of fermion, and a new type of fermion corresponding to each known type of boson. These partner particles had already been proposed as part of a theory called Supersymmetry in the late 1960s. The names of the new fermions were created by adding the suffix 'ino' to the names of their existing boson counterparts: so, there would be photinos, winos (for the W), zinos (for the Z), gluinos (for the gluon) and, crucially, gravitinos. The new bosons were named by prefixing the existing fermions with an 's': so, there would be squark, a selectron, a sneutrino. The new string idea, incorporating Supersymmetry, was called Superstring Theory.

Supersymmetry also held an aesthetic appeal for physicists who were studying the Standard Model. And so, the idea was carried across to the more traditional view of particles - as pointlike objects. The new scheme was dubbed 'Supergravity' and, thanks to the proposed existence of the gravitons and gravitinos, appeared to be very successful at overcoming the problems of the infinities associated with the pre-supersymmetry Standard Model. This was achieved in a development of Supergravity, called Extended Supergravity – a set of Quantum Field Theories that involved calculations in a number of extra dimensions. The most comprehensive was 'N=8 Supergravity', which was formulated in eleven dimensions. Our experience of the Universe, in three dimensions of space and one of time, might seem to be at odds with this theory. But it need not be. As long ago as 1919, German mathematician Theodor Kaluza (1885 - 1954) suggested that there may be more dimensions to the Universe than meet the eye. Kaluza proposed the existence of a fifth dimension. This simple but revolutionary idea allowed him to express both Einstein's General Relativity and Maxwell's Electromagnetism in a single theory. In Kaluza's plan, electromagnetism was expressed as curvature in the fifth dimension, in exactly the same way as gravity is explained as curvature in four dimensions under General Relativity.

Kaluza's idea was not greeted with much enthusiasm, despite support from Einstein, because there was no evidence of a fifth dimension. In 1926, however, Swedish physicist Oskar Klein (1894 - 1977) suggested a way in which it could be reconciled with observation and experience. Klein asserted that the fifth dimension could be curled up tightly, or 'compactified'. According to this idea, the extra dimension would form a tiny loop at every point in four-dimensional spacetime. Moving an object through ordinary, 'extended' spacetime dimensions would necessarily pass through the entirety of the fifth dimension many times. Mathematically, this undertaking worked, but the Kaluza-Klein hypothesis was not widely accepted as a description of the Universe. However, it was revived in the light of supersymmetry theories, which required the existence of several unseen dimensions. The original, bosonic, string theory involved vibrations of the strings in 26 dimensions. The various theories of Extended Supergravity also required the existence of several extra dimensions. Compactification provides a convenient and mathematically consistent method of 'hiding' the existence of these extra dimensions, and therefore of saving several promising ideas.

From its birth in the late 1960s, String Theory had only very limited support, largely because it had failed in its original assignment of describing the strong interaction. In addition, however promising the theory seemed to its adherents, it was plagued with incompleteness and inconsistencies. But during the 1980s, Superstring Theory enjoyed widespread acceptance, when its proponents managed to overcome some of its inherent difficulties. It was also shown that it contains all of the physics described by the Standard Model. All of a sudden, superstrings took centre stage in physics, and the theory was accepted as a serious contender for a theory of Quantum Gravity. In the 1990s, String Theory presented physicists with an embarrassment of riches: there were five different theories, each one with its own strengths and weaknesses. All the competing string theories involve vibrating strings at the Planck scale; all include the concept of supersymmetry; and all are formulated in ten dimensions (nine space and one time). Clearly, the four dimensions that we experience are accounted for, but what about the other six? Compactification of these dimensions is a possibility, and could have a physical meaning: perhaps gravity curved these dimensions extremely tightly in the very early Universe, and they have never 'unfurled'. Whatever the explanation of compactified dimensions, string theorists make use of mathematics developed by Eugenio Calabi (born 1923) and Shing Tung Yau (born 1949) to describe their configurations. Multi-dimensional shapes called Calabi-Yau manifolds may be the form of the hidden dimensions. If the extra dimensions posited by Superstring Theory really are compactified to a size comparable to the Planck scale, how can we ever find evidence of them? Well, Superstring mathematics predicts that the strength of the gravitational interaction would be affected at distance scales about the same as the compactified dimensions. No test has been carried out on the gravitational interaction at scales less than about a millimetre, although several are planned. However, the effect of enhanced gravitation may be detectable in experiments in extremely powerful particle accelerators. In a high-energy collision, graviton strings would 'wind' around the extra dimensions, and this would increase the strength of the interaction. Alternatively, a graviton created in a particle collision may be lost to other dimensions, resulting in a loss of energy that could be detected by measuring the energy of the particles present after the collision. So far, particle physics have found no evidence of the compactified dimensions, but the search continues.

Some of the five separate Superstring theories include the possibility that strings are not the only vibrating objects that can describe fundamental particles. Two dimensional membranes are also possible, as are other 'surfaces' in higher dimensions. In general, these vibrating surfaces are called p-branes. A (one-dimensional) string is called a one-brane, a two-dimensional surface is called a two-brane, and so on. Strings are likely to be most important, because two- or three-branes, for example, require much more energy for their creation. Superstring theory may also include an explanation of black holes, by considering objects called D-branes. In the Superstring theories that include open strings, rather than strings that form closed loops, there is an additional condition which the strings must satisfy. The ends of strings must attach to a multi-dimensional surface. That surface is a D-brane, a special case of p-brane. String theorists have found a way to explain a black hole as a collection of D-branes, and within their analysis, they worked out a way to calculate a black hole's entropy. As explained in Chapter Three, the entropy of a black hole is a measure of the black hole's disorder. But according to General Relativity, a black hole is a simple object that can have no internal structure or disorder. Stephen Hawking proposed that black holes might radiate with a spectrum characteristic of its temperature, but could not provide a way to calculate their entropy. If the D-brane approach to black hole entropy can be as successful as it promises to be, Superstring theory will receive a significant boost.

The search for a final theory In addition to the similarities between the five separate Superstring theories mentioned above, all of them have something else, more profound, in common: they are all linked by mathematical relationships called dualities. In 1994, String Theory guru Ed Witten (born 1951) interpreted these dualities as signifying that all five ten-dimensional Superstring theories are, in fact, approximations to a grander theory, which has been called 'M-theory'. Just as relativity theory was able to explain electromagnetism by going from three dimensions to four, and Kaluza-Klein theory was able to unite electromagnetism and gravity by going from four dimensions to five, M-theory is a Superstring theory in eleven dimensions, rather than the ten dimensions of the existing Superstring theories. The details of M-theory are, as yet, unknown, but most string theorists believe that it holds great promise for being, or at least developing into, a theory of everything.

An alternative to compactification of all the hidden dimensions is that some of the extra dimensions may be extended, but that all of our observations are restricted to just the four with which we are familiar. Along with this scenario comes a different explanation for the fact that the gravitational interaction is much weaker than the electromagnetic, weak or strong interactions at normal scales. In this case, we live on a three- or four-dimensional 'sheet', called a brane (this is not the same as the p-branes of Superstring theory). Gravity would be weaker than the other interactions because, being related to spacetime curvature at large distances, it would be able to 'leak out' to other dimensions. The kind of mind-blowing proposals that come from Superstring theory and brane worlds lie in the realm of cosmology, to which we turn in Chapter Six.

6: IN THE BEGINNING

Questions regarding the origin of the Universe are profound, and their ultimate answers may be forever beyond our grasp. Throughout most of human history, such questions have remained deeply rooted in the realm of speculative philosophy, mythology and religion. But in the past century or so, thanks to incredible advances in astronomy and in experimental and theoretical physics, it has been possible to address them scientifically, and to move towards answers. Cosmologists have constructed a scenario that describes the development of the Universe from its very earliest moments right up to the present day, based on well-tested physical theories – in particular, General Relativity and the Standard Model of particles and interactions. The story told by this scenario is more detailed than any mythological or religious description, and is extremely well-supported by a host of rigorous observations. Progress in cosmology during the rest of the 21st century will depend upon the successful unification of the theories of physics as well as continuing progress in astronomical observation. Science may be able to unlock the deepest secrets of the Universe after all – but many big questions still remain unanswered.

Expanding Universe Inside huge molecular clouds in space, gas and dust clumps together to form stars, as a result of mutual gravitational attraction – caused by the warping of space, as explained by General Relativity. Without the buoying up effect of the heating caused by nuclear fusion sizzling in their cores, stars would continue their collapse, as gravity is persistent in its quest to pull everything together.

With this in mind, you can understand why, in 1917, Albert Einstein proposed that there must be some kind of repulsive force counteracting gravity throughout space: otherwise, the matter in the Universe, and spacetime itself, would have collapsed in on itself. Einstein's repulsive force would be a kind of negative pressure produced by empty space. His suggestion was based on several assumptions – crucially, the idea that the Universe is static, neither expanding nor contracting. Einstein's proposed repulsive force, or 'cosmological constant', as it is called, was actually a term arising quite naturally from calculations in General Relativity; to mathematicians it is a 'constant of integration'. Einstein was wondering what General Relativity could tell us about the curvature of the spacetime of the Universe as a whole, rather than small regions of it. In 1919, Dutch astronomer and physicist Willem de Sitter (1872 - 1934) realised that Einstein's assumption that the Universe is static was unjustified. De Sitter, who had been studying the cosmological consequences of General Relativity ever since the theory was published in 1916, worked out solutions to Einstein's field equations in which spacetime could be expanding. But in these simplified solutions, there was no mass-energy present. If the Universe was expanding but had zero density, it would continue to expand forever.

During the 1920s, Russian mathematician Aleksandr Friedmann (1888 - 1925) produced a set of solutions to Einstein's equations that were based on a more realistic set of assumptions than either Einstein's or de Sitter's: the Universe contains mass-energy, is isotropic and homogeneous, and yet could be expanding. As explained in Chapter One, the Universe does indeed look the same in all directions, and does appear to have a uniform composition throughout, at least when considered on a large enough scale. Friedmann worked out how the fate of an expanding Universe depends upon the overall density of mass-energy and the rate at which the Universe is expanding. If the Universe is not expanding at all, then even a very low density of mass-energy will ultimately cause it to collapse on itself, since gravity has an infinite range. But even an expanding Universe may collapse: if the density is greater than a certain, critical value, there will be enough mutual gravity to slow the expansion, reverse it and pull everything together to a point. If the density is below the critical value, the Universe will expand forever at a more-or-less constant rate; if the density happens to be exactly the critical value, it will expand forever, but at an ever-decreasing rate. There are similarities between these ideas and the concept of escape velocity, discussed in Chapter Three. If you throw an object upwards from the surface of a planet, the object's fate depends upon two factors: how fast you throw it and the strength of the gravitational field of the planet. A high-density Friedmann Universe corresponds to a planet with a strong gravitational field. In such a Universe, the expansion would have to be relatively rapid if the Universe is to avoid ultimate collapse.

Friedmann's solutions showed how the fate of the Universe depends upon its current mass-energy density and its current rate of expansion. But they also made it possible to consider the curvature of spacetime on a cosmic scale. In a Universe with more than the critical density, spacetime will be positively curved. Such a Universe is said to be closed, and is finite in extent. A Universe with less than critical density has negative curvature, and is open and infinite. If the density is exactly equal to the critical density, the spacetime of the Universe is flat and infinite. In 1927, Belgian priest-turned-astrophysicist Georges Lemaître (1894 - 1966) independently developed solutions similar to Friedmann's, but he went further. As a result of his religious leanings, Lemaître proposed that an expanding Universe must have had a beginning. Few cosmologists had seriously considered this before, but Lemaître envisioned that the Universe might have originated in what he called a 'primeval atom' or 'cosmic egg': a tiny particle that somehow exploded long ago to form the expanding Universe. One important feature of Lemaître's scenario is that, given the current expansion rate and the current mass-energy density of the Universe, cosmologists would be able to calculate how much time has passed since the explosion of Lemaître's cosmic egg: the age of the Universe.

By the way, you might wonder what meaning time has on a cosmic scale, when Einstein's theories of relativity demand that time is a relative concept. Fortunately, in a homogeneous Universe, expanding at the same rate throughout, it is possible to define a 'cosmic time', because any distortions of time are the result of local gravitational fields or local motions. The expansion can be seen as a 'flow' and, on average, objects will be carried along with the flow, their local motions and gravitational fields becoming insignificant on a cosmic scale. Any galaxy moving with the flow keeps the same cosmic time as we do.

Back to the beginning Around the same time as the early cosmological models were being considered, astronomers were beginning to amass evidence that Friedmann, Lemaître and de Sitter were right: the Universe really is expanding. Because this means that the Universe is not static after all, Einstein abandoned his cosmological constant (he later called the idea the biggest blunder of his life) and also accepted the idea that there must have been a beginning. The evidence in favour of the expansion of the Universe was provided by measuring the redshift of light from distant galaxies. This is 'cosmological redshift', which Friedmann had predicted in the analysis of his cosmological models. As explained in Chapter One, cosmological redshift is a decrease in the frequencies of electromagnetic radiation emitted by a distant object caused by the expansion of the space between that object and the observer. In 1929, American astronomer Edwin Hubble analysed Vesto Slipher's observation of the redshifts of 40 galaxies, and estimated the distances to 24 of them. As a result, he derived 'Hubble's Law', discussed in Chapter One, which shows that there is a linear relationship between the distance of a galaxy and the speed at which it is receding. In other words, measure the speed of a galaxy at a given distance, then measure the speed of a galaxy twice as far away, and the second speed will be twice the first.

Hubble's law seems to hold for all galaxies, in all directions. It makes it possible to work out the rate at which the Universe is expanding. From his first set of data, Hubble made an estimate of the 'Hubble Constant', the number you obtain by dividing the speed of a galaxy by its distance from us. The Hubble Constant is normally expressed in units of 'kilometres per second per megaparsec'. Here, 'kilometres per second' is the speed and 'megaparsec' is a measurement of distance, equal to about 3.26 million light years. Hubble's estimated value was 464 kilometres per second per megaparsec: so, a galaxy 1 megaparsec distant would be moving away at 464 kilometres per second, while a galaxy at 2 megaparsecs would be moving away at 928 kilometres per second. Hubble's estimate for the Hubble Constant was wildly inaccurate, for three reasons. Firstly, he only had measurements relating to 24 galaxies, a very small sample when you consider how many galaxies there are altogether. Secondly, the method he used to calculate the distances to the galaxies was flawed. And finally, Hubble was only able to measure distances to galaxies that are relatively nearby; most of these are part of the same cluster or supercluster as our own galaxy, and are therefore subject to local motions that disrupt the linear pattern. Nonetheless, Hubble did manage to show that, as a rule, the further away a galaxy is, the faster it is receding – quite an achievement in itself. To refine the value of the Hubble Constant has been a major goal of astronomers, and the variation in measured values has often been a source of disagreement between them. One of the main objectives of the Hubble Space Telescope project was to derive an accurate, unquestionable value of this important number. The current best value is very different from Hubble's: just under 72 kilometres per second per megaparsec. So, a galaxy 1 megaparsec away will actually be moving more like 72 kilometres per second, not 464.

The dating game The value of the Hubble Constant is directly related to the Universe's rate of expansion. By mathematically running cosmic time backwards, cosmologists can therefore use it to work out roughly when the expansion began, and provide an age for the Universe. If Hubble's original figure of 464 was accurate, the Universe would be no more than about two billion years old – less than half the age of Earth. A high-valued Hubble Constant implies that the Universe must be young, since expansion must be rapid in the early stages to 'escape' the mutual gravity that would make it quickly collapse. Gravity slows the expansion over time, so the much reduced, and much more accurate, figure of around 70 kilometres per second per megaparsec gives a much longer history of between 8 and 15 billion years. The discrepancy between these figures is due to the range of possible values of mass-energy density present in the Universe – and the fact that the rate expansion has probably not remained constant. As we shall see, cosmologists now have a much better idea of the values of the mass-energy density and the way in which the expansion has changed, and can work out the age of the Universe with much greater precision than this.

In 1948, Russian-born physicist George Gamow (1904 - 1968) applied the theories of particle physics to Lemaître's idea, providing insight into what the young Universe might have been like. Together with his student, American physicist Ralph Alpher (1921 - 2007), Gamow worked out that the early Universe would have been an incredibly tiny, hot, dense soup of subatomic particles. Gamow and Alpher also suggested how this cosmic soup could have formed atoms of hydrogen and helium as the Universe expanded and cooled. They realised that a huge amount of radiation would have been produced in the energetic young Universe, and that it would have been characteristic of the extremely high temperatures in which it was created. Crucially, their theory described how the radiation would be constantly absorbed and re-radiated by the energetic particles in the first few hundred thousand years, but that at a certain time, when electrons joined nuclei to form atoms, radiation would be 'free' to travel across the Universe forever. This is the epoch of recombination, mentioned in Chapter One.

Gamow and Alpher even suggested that this thermal radiation would still be present in the Universe, hugely redshifted and therefore representing a much cooler temperature: a few degrees above absolute zero. Gamow's story was appealing, and it caught the imaginations of many cosmologists, though not every one. Uncomfortable with the idea that the Universe had a beginning – largely because that seemed to imply the existence of a creator – three other scientists produced an alternative scenario, also in 1948. They were British physicist Fred Hoyle (1915 - 2001) and Austrian physicists Hermann Bondi (1919 - 2005) and Thomas Gold (1920 - 2004). Their hypothesis was called the Steady State Theory. Convinced that the Universe had to be the same at all times as well as all locations, they proposed that matter is constantly being created from the energy of empty space as the Universe expands. Intending to dismiss Gamow's ideas, Hoyle referred to the concept of a hot, dense, rapidly expanding young Universe with an explosive beginning as “The Big Bang”. The name stuck.

The supporters of the Steady State Theory were keen to point out that there was no sign of any 'fossil' relics of the early stages of a Universe, a fact that supported their idea that the Universe was unchanging in time. However, in 1964, quite by chance, two physicists – German-born American Arno Penzias (born 1933) and American Robert Wilson (born 1936) – detected a faint microwave signal coming from every direction of the sky. The spectrum of the microwave signal had exactly the profile of thermal radiation of a few degrees above absolute zero. Their finding, analysed by another American physicist, Robert Dicke (1916 - 1997), was quickly identified as the fossil radiation from the early Universe. What Penzias and Wilson had detected is the afterglow of the hot, young Universe – the cosmic microwave background (CMB). It provided almost irrefutable evidence that the Big Bang hypothesis is correct. Remarkably, its temperature was within about one degree of Gamow's and Alpher's predicted value. With the discovery of the cosmic background radiation, the Big Bang theory finally became accepted by almost all cosmologists, and it seems extremely likely that it is correct. However, there are several features of the theory that were unexplainable: in particular, why space is expanding in the first place and where the original soup of subatomic particles came from.

In 1970, Stephen Hawking and Roger Penrose considered the development of the Universe in terms of black hole theory. As explained in Chapter Three, Penrose had shown that there must be a singularity (a point of infinite density) at the centre of any black hole. Picture the formation of a singularity in reverse, and you get an expansion, in which spacetime becomes less warped, and the density of mass-energy reduces continuously. The Universe must have begun in this way. Penrose and Hawking showed that there must be a singularity at the beginning of any Friedmann-type expanding Universe. The Big Bang singularity was not a black hole, however, despite the fact that all of the mass-energy of the entire Universe was contained within an infinitesimal volume. There are several reasons why the Big Bang was not a black hole, including the fact that spacetime was expanding. I shall shortly consider why and how the Universe began expanding; it must have been rapid to prevent the young Universe from collapsing to form a black hole. The origin of the singularity at the beginning of the Universe remains a mystery. It could be the collapsed final state of a previous Universe – leading to the idea of an 'oscillating' Universe that repeatedly expands and contracts. Alternatively, it could be a quantum fluctuation, occurring either in nothingness or from within another Universe. In that case, it could be that other Universes are being created all the time, as quantum fluctuations in our own.

The value of the Hubble Constant gives only a range of possible ages for the Universe, and on its own says nothing about its ultimate fate. To refine the estimate of the Universe's age, and to determine its fate, it is necessary to find its current density. There have long been good theoretical reasons why the value of the density must be close to the critical density, and recent observations support this proposal. If it was only slightly greater than critical, the Universe would have collapsed in on itself long ago. If it was much less, then the expansion would have been too rapid for galaxies, stars and planets to form. For convenience, cosmologists refer to the actual density by comparing it with the critical density, using the Greek letter Ω (omega). If the Universe has an average density exactly equal to critical density, Ω would be exactly 1.

So Ω is simply the ratio of the actual density of the Universe to the critical density. It is important to bear in mind that the value of the critical density (but not the actual density) depends upon the rate of expansion – the value of the Hubble Constant.

Fortunately, astronomers can make good estimates of the current density of mass-energy in the Universe – and therefore, armed also with an accurate measurement of the Hubble Constant, can work out the value of Ω. There are several methods by which they can achieve this. The most obvious method is to count how many galaxies are present in a given volume of space and assess the average mass-energy of each one. The figure must include the radiation produced by the galaxy as well as the mass of all the stars and interstellar gas and dust, since radiation, too, contributes mass-energy. However, as discussed in Chapter One, there is a large amount of mass-energy unaccounted for in such an estimate. Astrophysicists and cosmologists call this missing mass-energy 'dark matter', and its presence can be inferred from its gravitational effect on stars and galaxies. So, the best way to work out the cosmic density is to observe the speeds and directions of motion of galaxies in clusters or superclusters of galaxies. In that way, cosmologists can include an estimate of the amount of dark matter. It seems that dark matter must make up about 95 percent of the mass-energy that can be detected in this way. As explained in Chapter One, astronomers and cosmologists are sure that dark matter must exist, but they are unsure of its nature. Unfortunately, the amount of mass-energy that astronomers calculate as being present, even including the dark matter that they know must be there, falls well short of the figure needed to make the value of Ω anywhere near equal to 1. The shortfall is significant: the best estimates put the figure at about 0.3 - that is, 30 percent of the critical density, given the value of the Hubble Constant.

The discrepancy between the observed current value of Ω and what theory demands, since galaxies, stars and planets do exist, is worryingly large. But extrapolate back to the earliest moments in the Universe's history, and the problem blows up to phenomenal proportions. The critical density of the Universe changes with time, just as the actual density and the Hubble Constant do. It turns out that, for the Universe to look anything like it does today (to be flat) the mass-energy density present in the earliest fraction of the second must have been finely tuned to one part in many trillions. How could our Universe have been so delicately balanced between a short-lived damp squib and a rapidly expanding cold, empty and featureless spacetime arena? This is what cosmologists call the 'flatness problem'. It was a major stumbling block for cosmology in the 1960s and 1970s. The other main problem with the original Big Bang hypothesis is the so-called 'horizon problem'. The Universe is extremely isotropic and homogeneous on the largest of scales. The cosmic microwave background gives perhaps the best example. The profile of the microwave radiation is an exact fit to that of an object at a temperature of 2.72548±0.00057 degrees Celsius above absolute zero, and the variations between different regions of the sky is less than one part in ten thousand. This presents a problem to cosmologists, because such uniformity suggests that the various regions must have been 'causally connected'. To see what this means, imagine heating one of two metal objects that are in contact. If they remain in contact for long enough, the temperature will be the same throughout each object, and each object will have the same temperature: they will be in thermal equilibrium. If you move the two pieces of metal apart before they have had time to reach equilibrium, they will forever be different temperatures. The early Universe was expanding so quickly that regions separated by anything more than about a one degree angle in the sky would not be causally connected, even though the various regions would be connected by radiation travelling at the speed of light.

The benefits of inflation The flatness problem and the horizon problem were serious issues that tainted the success of the Big Bang scenario. But in 1981, American particle physicist and cosmologist Alan Guth (born 1947) put forward an inspired theory that solved both problems at a stroke. The

Cosmic Inflation Model, as this is called, involves a brief period of incredibly rapid expansion beginning 10^{-35} seconds after the start point of the Big Bang. This inflation would have lasted last for just 10^{-32} seconds, but during that time, the Universe would grow by a factor of 10^{30} or more. This is equivalent to an atom growing to one light year diameter in less than the time it takes for light to travel a distance equal to the diameter of a proton!However, the Universe would have been far smaller than an atom, or even a proton, at the beginning of inflation. Other scientists had suggested that inflation would cure the Big Bang theory's ills, but they could not suggest a mechanism by which inflation could occur. What could provide the energy for such a rapid expansion? Guth's background in particle physics – his understanding of the Standard Model of particles and interactions, and in particular the idea of symmetry and Grand Unified Theories described in Chapter Five – enabled him to find just such a mechanism, with just such energy.

In his inflationary scenario, Alan Guth appealed to the properties of a certain type of field in Quantum Field Theory, called a scalar field. Examples of scalar fields are those involved in the Grand Unified Theories that describe the weak, strong and electromagnetic fields in a single, unified framework. Another is the Higgs field, responsible for endowing fundamental particles with mass. The important feature of scalar fields is that they have their minimum energy when their strength is a certain value above zero, not when it is at zero. In other words, what we think of as a vacuum, in which the value of all fields is zero, has more energy than spacetime filled with scalar fields of a certain value. So, when the Universe was born, there was a 'false vacuum', whose natural tendency is to become a 'true vacuum', releasing energy. A false vacuum has a high density of energy, which cannot quickly be reduced. It also exerts negative pressure, which in the inflation would have caused the Universe to expand rapidly. The energy released had the effect of producing a hot soup of fundamental particles.

As the field moves from a false vacuum to a true vacuum, it undergoes what physicists call a phase transition, not unlike what happens when water turns to steam. And just as steam forms as growing bubbles inside a body of water, true vacuum would grow as 'bubbles' inside the false vacuum of the field. The edges, or walls, of the bubbles would form into large scale structure of the Universe today; clusters and superclusters of galaxies are indeed observed along the walls of enormous bubbles of apparently empty space. The original theory of inflation predicts that these large scale structure should be more dense than they are observed to be in the real Universe; in other words, it is wrong. Several variations of cosmic inflation have been put forward since the 1980s, however, which address this and other problems with the original theory. In 'new inflation' and 'chaotic inflation', the hypothetical scalar field that is responsible for inflation is called the inflaton field. This field is still a false vacuum, but its strength is unaffected by inflation; it continues to exist, and other Universes can emerge from it, as quantum fluctuations. At the end of the inflationary period, the Universe would have been about 10^{-30} seconds old, the size of a grape, and awash with X bosons (leptoquarks), which would quickly decay to form leptons and quarks. The results of the brief epoch of inflation, then, produces just what is needed in the starting point of the Big Bang scenario: a small, expanding Universe containing leptons and quarks. The real strength of the inflation theory is that it explains the current state of the Universe: flat and homogeneous. The Universe is smooth and flat because inflation is like pulling a crumpled sheet rapidly from all corners. It is homogeneous, because regions on opposite sides of the sky were causally connected for a brief moment before inflation.

While the inflationary scenario provides a plausible explanation of the evolution of our Universe from its very earliest moments until now, it does not extend back to time zero. It assumes the existence of a starting point: a tiny Universe ready to inflate. The most promising attempt to work out what happened before this is Quantum Cosmology. One of the main aims of Quantum Cosmology is to account for –

or preferably to avoid the need for – the singularity that appears in any standard Big Bang model, including the inflationary one. Quantum Cosmology is the application of the principles and procedures of quantum physics to the Universe as a whole. For a complete description of the events prior to inflation, physicists would need to appeal to a successful theory of Quantum Gravity, which is not yet forthcoming. Nonetheless, quantum physics can provide useful insight into the ultimate origin of the Universe. Just as systems of particles can be in certain 'quantum states' described by a wavefunction, some quantum cosmologists are keen to determine the 'wavefunction of the Universe'. The 'state of the Universe' in Quantum Cosmology is very simplified, often so much so that it describes only the size of Universe together with an overview of the matter fields within it. It is then possible to apply quantum mechanics to this simple picture. The result is a wave equation called the Wheeler-De Witt Equation, which describes the development of the Universe in quantum terms.

In quantum physics, Richard Feynman's 'sum over histories' approach (see Chapter Four) provides one way of interpreting the outcome of experiments involving particles. It allows physicists to investigate the outcome of an experiment in terms of the probabilities of every possible sequence of events (history) that led to that outcome. In Quantum Cosmology, the 'outcome' is the current state of the Universe. By knowing the initial state of the Universe, and all possible histories through which it can evolve, it is possible to calculate the probability of any particular state of the Universe at a later cosmic time. The final state and the initial state of the Universe in this analysis are called boundary conditions. The initial state of the Universe is exactly what cosmologists want to know, so on the face of it, the sum over histories approach appears useless. However, in 1983, Stephen Hawking and American physicist Jim Hartle (born 1939) formulated the 'no boundary proposal', which asserts that the Universe had no initial state: it was created from nothing. A sum over histories for the Universe is an exercise in four-dimensional analysis, because cosmic time is treated on an equal footing as the three space dimensions. So, the state of the Universe at any moment of cosmic time will be a three-dimensional slice through any of the four-dimensional shapes that describe its possible histories. The possible histories of the Universe will include all those through which some three-dimensional slice can represent the current state of the Universe. Some of these possible histories can be effectively ruled out as having minuscule probabilities, and quantum cosmologists concentrate only on the most likely histories, called 'instantons'.

In each instanton, the origin of the Universe lies in a region of four-dimensional space in which no particular dimension is cosmic time. In this state, time has no meaning. The pre-inflation Universe can be reached from an instanton by 'quantum tunnelling'. This is a common feature of well-known microscopic systems, and is due to the probabilistic nature of the world, as described by quantum physics. Quantum tunnelling is the cause of alpha decay, for example, in which an alpha particle (two protons and two neutrons) escapes from a nucleus. Quantum physicists can calculate the probability that an alpha particle can 'leap' across the apparent barrier that should happily confine it. One consequence of tunnelling out of an instanton is that an infinite number of Universes might be produced from such a four-dimensional region.

Quantum Cosmology is an effort to probe what might have happened before cosmic inflation. Many of the models it has produced place restrictions on the precise appearance of the cosmic background radiation. So, these restrictions can be thought of as predictions, and this means that quantum cosmological models can effectively be tested by examining the radiation that makes up the cosmic microwave background.

In the background The discovery of the cosmic microwave background in 1964 heralded a new beginning in observational cosmology. The microwave radiation that bathes the Universe today is like a frozen snapshot of the exact conditions in the Universe when it was about 300,000 years old. We cannot see further back in time, because before the epoch of recombination, electrons were free from nuclei, and free electrons scatter electromagnetic radiation as soon as they absorb it. This made the early Universe totally opaque; after the electrons had become bound to atoms, space became transparent. The cosmic background radiation represents a 'last scattering surface'. It provides observational evidence to support the Inflationary Big Bang theory, as well as opportunities to test the predictions of sophisticated new variants of it, produced by Quantum Cosmology.

The extreme uniformity of the cosmic microwave background has enabled astronomers to determine the 'local' motion of the Earth relative to the overall expansion of space: the Hubble flow. Objects that are simply carried along with the expansion of space have no such local motion. The Earth, however, is moving around the Sun; the Sun is moving around the Milky Way Galaxy; and the galaxy may well be moving relative to the Hubble flow, pulled this way or that by gravitational interaction with other galaxies in the Local Group. These movements, if fast enough, should show up as variations in the frequency of the cosmic background radiation, caused by the Doppler effect. In 1977, several experiments did indeed detect a very slight difference in frequency between one side of the sky and the other. When the results of these experiments were adjusted to take into account the motion of the Earth around the Sun, the Sun around the galaxy and the galaxy in our Local Group, astronomers determined that even the Local Group itself has a local motion of more than 600 kilometres per second relative to the Hubble flow. This is explained by the fact that the Local Group is itself part of a vast supercluster of galaxies about 100 megaparsecs in diameter. The combined mass-energy of the entire supercluster, including its dark matter, is causing the Local Group's local motion.

The importance of the cosmic microwave background does not stop at giving astronomers a way to measure the motion of our group of galaxies, however. Detecting slight variations in the temperature of the radiation provides a wealth of information about conditions in the Universe at the epoch of recombination. It also gives cosmologists several vital ways to test their theories. At first, there seemed to be no variation whatsoever in the temperature of the background radiation, apart from that caused by our local motion with respect to the Hubble flow. In one sense, this was a triumph for the Inflationary Big Bang scenario: isotropy and homogeneity are important consequences of the story it tells. However, if the early Universe had been perfectly smooth, then it still would be. Only if there were slight variations – inhomogeneities – in density in the early Universe would there have been stars, galaxies and the large scale structure of clusters and superclusters. Cosmic Inflation theory predicts the existence of irregularities as quantum fluctuations in the inflaton field. These inhomogeneities in the early Universe, tiny ripples in the fabric of spacetime, would have to show up as 'anisotropies' in the background radiation.

In one of the landmark achievements of modern cosmology, an orbiting microwave telescope called COBE (the Cosmic Background Explorer) detected anisotropies in the cosmic background radiation, in 1992. Even so, COBE's instruments had only low resolution: they could only observe variations on an angular scale of about 10 degrees. The variations detected by COBE correspond to structures about 1000 megaparsecs in diameter in today's Universe – more than thirty thousand times the diameter of our galaxy, and even bigger than the diameter of the supercluster of which we are a part. There must have been smaller inhomogeneities than these in the young Universe. Indeed, inflation theory predicts that there will be other variations, on various angular scales. The most important should be on the 1 degree scale. Regions in the early Universe that were causally connected before the epoch of recombination should be fairly uniform, and these would appear in the background radiation, as granulations about one degree across. Since the late 1990s, several experiments have detected these one degree scale patches. The angular size of these uniform regions provides cosmologists with another vital piece of information: it confirms that the Universe is indeed flat or very close to flat (so, Ω is 1 or very close to 1). If the spacetime of the Universe on cosmic scales had positive curvature, these regions would appear larger than one degree;

if it had negative curvature, they would appear smaller than one degree.

Further regions, with smaller angular diameters, are predicted and have also been detected. Each region appears as a peak in a graph called a power spectrum, which shows up regular variations in the temperature of the background radiation. The first peak is the strongest, and corresponds to the existence of the one degree patches. From the height of the second peak relative to the first, cosmologists can calculate how many baryons (protons and neutrons) were present in the early Universe. This helps them to work out whether dark matter is 'baryonic' (ordinary) or non-baryonic. It turns out that the inflationary model is accurate in its predictions of the amounts of baryonic matter, and that astronomers' observations of the density of baryonic matter in the Universe today are just about right. The peaks in the background radiation power spectrum actually correspond to sound waves in the Universe frozen in time at the epoch of recombination. There is no sound in space in today's Universe, because sound needs matter through which to propagate. But the Universe was much more dense at the epoch of recombination, and sound waves – caused by oscillations of matter due to gravitational instabilities – would have existed. Current research into the cosmic background radiation is therefore focussed on the acoustics of the Universe at the epoch of recombination, and what they can tell us about conditions in the early Universe. Current and future projects aim to produce increasingly precise plots of the power spectrum.

Analysis of the cosmic background radiation has produced a consistent set of results that are in line with inflationary cosmology. From its launch in 2001 until the end of its observational run in 2010, NASA's Wilkinson Microwave Anisotropy Probe (WMAP) provided an extremely clear view of the early Universe. From the results, it was possible to put very strict limits on some key cosmological parameters, including Ω (the Universe's density), the Hubble Constant and the age of the Universe. A sister to WMAP was Planck, a space observatory that took measurements between 2009 and 2013, gleaned results in agreement with WMAP's with even higher levels of accuracy. All of these observations give tremendous support to the theory of the Big Bang with cosmic inflation.

However remarkable these results are, there still remain a few major problems with modern cosmology. Only about 5% of the mass-energy can be accounted for in ordinary matter: galaxies and stars. Dark matter 'detected' by its influence on the motions of galaxies can only contribute a maximum of about 27%, and even the nature of that matter is unknown. The remaining 68% or so of the mass-energy necessary to make the Universe flat is an even greater mystery. There is one distinct possibility, however: there may be a field throughout space that is contributing the majority of the mass-energy. There is even evidence that the energy of this field is making the Universe expand at an increasing rate. It might just be that Einstein's cosmological constant idea was not a blunder after all.

Faster now Cosmology since the discovery of Hubble's Law has been based on the assumption that Einstein's 'anti-gravity' cosmological constant was nonexistent. The fate of our expanding Universe predicted in standard cosmological theories, with no cosmological constant, is simple to predict, once the values for the Hubble Constant and Ω are known. The rate of expansion in this simple picture continuously slows down due to gravity, and the value of Ω determines the rate of that slowdown. If the slowdown is rapid enough, the expansion will reverse and the Universe will collapse. If the cosmological constant does exist, however, the picture is not so simple: a repulsive force would speed up the expansion, or at least affect the rate at which the expansion slows. While the existence of a cosmological constant is an unwelcome complication to the overall picture, its energy could account for the missing 68% of the mass-energy necessary to give the Universe exactly critical density. The cosmological constant would have the same value, and therefore the same energy density, throughout space, and so would not affect the local motions of the galaxies. It would therefore not have been detected in calculations of the amount of dark matter in the Universe.

In 1998, astronomers found evidence that the Hubble Constant has increased over time; the expansion of the Universe really is accelerating. So, there must be some kind of repulsive force after all. The nature of the repulsive agent is not known; cosmologists refer to it as 'dark energy'. It could be that dark energy is simply the energy of empty space, which results from virtual particles popping into and out of existence as quantum fluctuations. The fact that there is energy in a vacuum has been demonstrated in the Casimir effect, a tiny force exerted on objects extremely close together, but not touching, in a vacuum. The force is derived from the fact that the lowest energy states of any quantised energy fields are not zero, as discussed in Chapter Four. In principle, the energy of the vacuum is a very likely candidate for dark energy: it should produce a repulsive force (a negative pressure), it is evenly distributed throughout space, and it would contribute greatly to the mass-energy of the Universe, as required. Unfortunately, attempts to calculate the density of the vacuum energy based on the Standard Model of particles and interactions give values that are either embarrassingly large or zero. If the large values were correct, the Universe would be far too dense: trillions of times more than critical density. If the vacuum energy is zero, then of course, dark energy must be something else. As the Universe grows in size, the density of ordinary mass-energy and dark matter decreases with time, while the density of vacuum energy, and its repulsive force (negative pressure), remains the same. If dark energy is vacuum energy, then, it becomes more dominant as cosmic time passes, and the Universe will therefore expand forever at an ever-increasing rate.

Several other candidates for dark energy have been put forward by cosmologists and particle physicists keen to explain why the expansion of the Universe is accelerating, and why nearly 70% of the mass-energy density is unaccounted for. Some theories involve a hypothetical substance called quintessence, evenly distributed throughout space, and causing a negative pressure, just as the vacuum energy is thought to do. The word 'quintessence' is from the Ancient Greek idea of a fifth element – equivalent to the aether, discussed in Chapter Two. The quintessence would result from the existence of scalar fields, like the one that would have caused the Universe's initial inflation. What makes quintessence attractive to cosmologists is that its value would change over time. Whereas vacuum energy is less important in a young, high-density Universe, but increases in importance as the Universe expands, quintessence is more important in a young, dense Universe, but less important now. A varying cosmological constant, which is in effect what quintessence is, could provide better explanations of the structure in the Universe. If it was unchanging over time, like vacuum energy, then the Universe may well have expanded too quickly in its youth for stars and galaxies to form. Quintessence is hotly favoured by some cosmologists, while many others are wary of it. They feel that the existence of this strange repulsive, all-pervasive substance has been suggested *ad hoc*, simply to account for troubling observations. Of course, that is often how science makes its most impressive strides forwards – as long as new theories are testable, of course. One test for the nature of dark energy, and the existence of quintessence, may be found in more precise measurements of the peaks in the power spectrum of the cosmic background radiation.

The main problem with quintessence is that its effects have never been observed in any laboratory; if it exists, it should be a fundamental interaction, like the strong, weak and electromagnetic interactions. However, quintessence must be very weak over small distances, so perhaps the Universe itself is the only real laboratory available to look for it. The first step in this procedure is to gain more accurate knowledge of the extent to which the expansion of the Universe is accelerating. The quest to understand dark energy and the accelerating expansion of the Universe is likely to remain a major feature of 21st century cosmology. The most promising progress may come from particle physics: if a successful theory of Quantum Gravity is formulated, then the mechanisms behind the expansion of space will be better understood.

Cosmology with branes The theory of the Inflationary Big Bang is very successful at describing how the Universe came to look and behave as it does, and is based on extremely well-tested theories of matter, space and time. The scenario it describes matches many detailed observations of the Universe on a grand scale, including the nature of the cosmic background radiation. Since inflation relies on Grand Unified Theories, it is restricted in its scope to events that occurred after 10^{-43} seconds. This is the Planck time, explained in Chapter Five. Events on time scales less than this can only be described by a successful theory of Quantum Gravity. And, since all aspiring theories of Quantum Gravity involve supersymmetry and the existence of several extra dimensions, future developments in theoretical cosmology are likely also to have these features. Cosmologists have begun to turn to the most promising – Superstring Theory – to provide the answers. In the late 1990s, an idea from Superstring Theory hit the world of cosmology and caused a flurry of activity.

In the 1990s, string research underwent a revolution, when Ed Witten showed that the five existing Superstring theories seemed to be different facets of a unified scheme, which has been called M-theory. While the five separate theories were formulated in ten dimensions, M-theory is an eleven-dimensional theory, as is the theory of Supergravity. One reason why these theories require extra dimensions is that gravity is very much weaker than the other fundamental interactions; this is called the 'hierarchy problem'. In three dimensions of space, gravity obeys an inverse square law, so that objects twice as far apart will experience one quarter of the strength of a gravitational interaction. If there were more than three dimensions of space, gravity would appear much weaker: in four space dimensions, for example, gravity would obey a law involving the cubed value of the distance, so that objects twice as far apart would experience one eighth of the strength of a gravitational interaction. This raises an intriguing possibility: that gravity really is comparable in strength to the other interactions, but that there are more than three dimensions; the fact that we only experience three of the space dimensions means that we observe an inverse square law. Multi-dimensional theories explain the fact that we appear to live in a four-dimensional world, three space and one time, by suggesting that the unseen dimensions are curled up, or compactified, so tightly that we cannot observe them. But if gravity really is much stronger, with its true strength concealed from us by its excursion into the hidden dimensions, then it should be possible to observe an increase in the strength of gravity at the scale of the unseen dimensions. So, any experiments that find evidence of a change in the gravitational interaction at small scales could provide evidence of compactified dimensions. From high-energy experiments already carried out, it is known that, should they exist, the extra dimensions must be smaller than a millimetre or so in diameter.

In 1998, physicists Nima Arkani-Hamed (born 1972), Giorgi Dvali (born 1964), Ignatios Antoniadis (born 1955) and Savas Dimopoulos (born 1952), working at Stanford University, California, USA, came up with an alternative solution to the hierarchy problem. They suggested that the four-dimensional Universe we know and love could be a membrane, or 'brane' in a five-dimensional 'bulk'. The fifth dimension would be large, perhaps even infinite, and not compactified. A brane is a Superstring-inspired concept: it is a multi-dimensional surface at which the ends of 'open' strings must terminate. In the brane picture of the Universe, gravitons – the virtual boson carriers of the gravitational interaction – would be closed looped strings that could exist beyond the brane, in the bulk. All other interactions would be carried by open string bosons, confined to our brane. Brane world theories may even explain dark matter: our brane could be folded back on itself, so that what we think of as dark matter is ordinary mass-energy in our own Universe, affecting the galaxy clusters that we observe by jumping from one point in our brane to another across the fifth dimension. This theory had a problem, however: the energy of the brane would warp the fifth dimension, effectively focussing gravity back onto the brane. As a way around this, physicists Lisa Randall (born 1962) and Raman Sundrum (born 1964) suggested the existence of a second, 'shadow' brane close to our own.

In the Randall-Sundrum Theory, our Universe is a low-energy brane, and the neighbouring brane has far greater energy. The second brane therefore 'soaks up' most of the gravitons, leaving us with a diluted gravitational interaction as before. In the Randall-Sundrum Theory, dark matter may exist on the second brane, affected across the fifth dimension, via gravity. The theory may also shed some light on dark energy. In particular, it may tell cosmologists why the vacuum energy appears much smaller than the Standard Model predicts if it is to act like a cosmological constant: it, too, may leak across the fifth dimension and into the shadow brane.

Theories involving branes can also be applied to questions of the origin of the Universe. One of the more exotic and revolutionary ideas to come from this line of research is the ekpyrotic scenario, suggested in 2001 by Neil Turok in the UK and Burt Ovrut, Justin Khoury and Paul Steinhardt in the USA. The ekpyrotic scenario provides an alternative to the Inflationary Big Bang theory. While it agrees that the Universe has been expanding for 13.8 billion years, it tells a different tale about how and why the expansion started. Initially, the Universe is a cold, dark, flat four-dimensional brane that is free to travel in the five-dimensional bulk. It collides with another brane (13.8 million years ago), and the two branes stick together. The collision releases enough energy to produce matter and the expansion of the Universe. Interestingly, the collision happens simultaneously at all points on the branes, and produces ripples – gravitational waves – that seed the formation of inhomogeneities in the early Universe. These inhomogeneities become galaxies and galaxy clusters in the Universe we observe today. The word 'ekpyrosis' is a Greek word meaning 'conflagration'; an energetic fire or a violent conflict. The ekpyrotic scenario is more than the crazed musings of cosmologists gone mad, however. It addresses problems with the Inflationary Big Bang Theory. For example, there is no singularity in the new scheme, and no magnetic monopoles are created, because the temperature of the brane after the collision is not as high as that predicted in the inflationary scenario.

Whether or not it can be confirmed, the ekpyrotic scenario does highlight the fact that there may be other ideas that can rival the success of the Inflationary Big Bang Theory. The real understanding of the origins and the nature of our Universe will only become clear when and if a successful theory of Quantum Gravity is formulated.

One cunning theory concerning the origin of the Universe – one that involves quantum gravity – is Cosmological Natural Selection, proposed by American physicist Lee Smolin (born 1955). This theory is based on a requirement that arises when you examine cosmology on the basis of loop quantum gravity, discussed in Chapter Five. In loop quantum gravity, there is no singularity to worry about at the beginning of the Universe, because gravity is quantised. It turns out that loop quantum gravity predicts a kind of symmetry: every new universe must arise from the collapse of another, or at least the gravitational collapse of a region of spacetime, such as a black hole. So, new universes may be popping into existence from black holes in our Universe, and our Universe may itself have been spawned from a black hole in another universe. In this picture, any universe whose laws of nature permit the existence of stars, and therefore black holes, can give rise to a host of other universes; each 'generation' would have laws of nature similar to the one it came from. Only universes whose laws allowed for the production of stars would be able to 'spawn' new Universes. Although this idea is little more than speculation at present, it would explain why the laws of nature are so well tuned to producing galaxies, stars, planets and life.

Any comprehensive theory that successfully explains the origin of the Universe will also rely on a clear interpretation of how the complex mathematics of modern physics relates to what the Universe is really like. I shall discuss what physics can tell us about the nature of reality in Chapter Seven.

7: WHAT IS REALITY?

> "These are merely names of the images we substituted for the real objects which Nature will hide forever from our eyes. The true relations between these objects are the only reality we can ever obtain."
> – Henri Poincare, 1905.

You are a human being, gravitationally bound to a planet that is in orbit around an average-sized star, in the outskirts of a typical galaxy. Your galaxy is part of a cluster, which – like all the others – is being carried along with the 'flow' caused by the expansion of spacetime. On the largest scale, gravity (described by the warping of space-time) is working against the expansion, but it seems that another force, perhaps originating in empty space itself, may be accelerating it. The expansion seemed to begin 13.8 thousand million years ago, when the entire Universe was much, much smaller.

You are a human being, made of molecules, which in turn are made of atoms. There are about 10^{27} atoms in your body altogether. Every atom is composed of protons, neutrons and electrons. Every proton and every neutron is composed of quarks. The electrons and protons are constantly exchanging virtual photons, which keep each atom together via the electromagnetic force. The protons and neutrons are exchanging virtual pions, themselves made of quarks, which keep each nucleus together via the residual strong force. The quarks are exchanging virtual gluons, which keep them confined in threes as protons and neutrons, or in twos as pions. The quarks, leptons and virtual bosons (photons and gluons) can be described by wavefunctions, which describe the probabilities of those particles having certain properties or being in a certain position. The virtual bosons pop into existence and out again before they can be detected. Every particle is a quantum of energy in a field, and fields permeate all of spacetime.

The map is not the territory All of the facts in the paragraphs above are correct, according to the rigorously tested theories of science. It is amazing that so much detail can be put into this portrait of your existence (in fact, there is much more besides). But why *are* there fields? Why *is* there a Universe at all? Science cannot really answer these questions: it can only answer 'how', not 'why'; ultimately, it can only describe, never truly explain. Most people want more than this: they want to know what a field is 'really' made of; they want to know what atoms are 'really' like, and whether the fundamental constituents of matter are 'really' particles or waves. The answers to these questions can never be answered. They lie in the realm where physics meets philosophy.

Before 1905, scientists believed that space and time were absolute quantities, just as common sense tells us that they are, and that atoms are tiny, impenetrable balls existing in a void. That view, we now know, is not a true reflection of reality. That is not to say that science is a nonsense, or that the 19th century physicists were stupid. In fact, quite the contrary: it is one of the great strengths of science that scientists are willing to abandon long-held beliefs about the nature of reality in the face of contradictory evidence; and the 19th century scientists made essential steps along the road towards truth. Religions, on the other hand, extract their truths from 'received wisdom', by revelation, for example, and they tend to hold on to them tenaciously. (That is not to say that religion is a nonsense, or that religious people are stupid, either.)

The concrete explanations that potentially satisfy us can only ever be analogies, based on the qualities of the macroscopic world that we can relate to; they will always be incomplete. Atoms really do behave as tiny, impenetrable balls in many circumstances: that analogy is sufficient to explain many of the bulk properties of matter such as pressure and temperature, for example. But to understand atoms in greater depth, the analogy must be abandoned: atoms are not solid balls. Instead, they have complex inner structure, ultimately made of collections of fundamental particles, themselves made of interacting fields. So what about fields? If they are the ultimate units of reality, what are they made of? We can picture fields in terms of useful analogies – as bodies of water or as a region of space filled with vibrating particles, for example. But in essence, a field is really just a set of numbers. A field has a value at every point in spacetime; in a sense, fields define spacetime. For otherwise, what are space and time? Fields are not actually made of anything: everything is made of them: made of numbers. In fact, to some scientists today, reality is made of 'information', for fields carry information.

Science can only really tell us what reality is *not*. It can rule out explanations or analogies that are inconsistent with empirical or rational truths. Empirical truths are observations of reality; if theory is in conflict with well-conducted experiments, then it is clearly the theory that is wrong. Rational truths are those that are consistent with logic or mathematics; here, scientists must be more careful. For example, the prediction of electrons with negative energy, which arose from Dirac's equation (Chapter Four) seemed illogical. But rather than rule out this prediction out of hand, Dirac realised that it had some meaning: the negative energy states were describing real particles: positrons.

Quantum reality The rational world of logic is based on our experience of the world. It seemed logical, in the mechanistic framework of the 19th century, that the particles of matter would interact deterministically. In other words, if you repeat an experiment in every detail, then you should expect exactly the same result. And indeed, on the macroscopic scale, this seems true. Fire the same cannon ball many times, with exactly the same velocity, from exactly the same spot, in exactly the same weather conditions in each case, and it will follow the same trajectory, and land in the same location, every time. The result of the experiment is completely determined by the initial conditions, and the correct laws of physics can be used to predict exactly what will happen. This is part of the cosy interpretation of reality assumed by the mechanistic view – although, it is not so cosy in that it means every event is determined, perhaps even *predetermined*, by the initial conditions. Since even human beings are made of particles, which obey physical laws, this seems to outlaw the concept of free will. It suggests that we have no control over what we do – and yet we feel as though we can make decisions and influence events. Nevertheless, it is likely that free will is simply an illusion: that we behave according to deterministic laws, and that consciousness arises, or 'emerges' and makes sense of our actions after they have happened.

Quantum physics gives some comfort to those who are troubled by the supposed loss of free will in a deterministic Universe. In the quantum world of microscopic particles, an experiment repeated many times may have many potential outcomes; which of the possible outcomes is the actual one seems to be a matter of random chance. In fact, this is little comfort, because it shifts the 'responsibility' for determining our actions from strict, unflinching, deterministic laws to random chance. Besides, apart from this random element, the quantum world is still deterministic: the wavefunction, which carries all the information about a particle or a system, including the probabilities of each possible outcome, behaves completely according to deterministic laws. When applied to a large number of particles, the indeterminate quality fades away. This is similar to the situation with the particles of a gas. Choose one particle, and there is no way that you can predict what speed it will be moving at. But look at a thousand million particles in the gas, and you can predict their behaviour with complete confidence. This is also why the trajectory of the cannonball is so easy to determine. It is impossible to predict the exact trajectory of one of the atoms within the cannonball, but you can predict where the cannonball as a whole will end up.

However, quantum physics offers a little more comfort than simply stating that the outcome of a particular experiment is at the mercy of random chance. Inherent in the formulation of quantum physics – in fact, intrinsic to reality on the microscopic scale – is the Uncertainty Principle. The quantum world is, by its very nature, 'fuzzy' and undetermined. Furthermore, the act of measurement affects the outcome of an experiment. Quantum physics blurs the line between the observer and the observed. We are not simply 'in' the world: we are a dynamic part of it. Leave a quantum mechanical system – a radioactive nucleus, for example – unobserved, and quantum physics suggests that it remains in a state of superposition: it can actually be said to exist as a combination of all its possibilities. Once the system is observed, its wavefunction 'collapses' and its properties become well-defined. A radioactive nucleus, for example, has a random chance of decaying in any given amount of time, and that chance can be calculated accurately using Quantum Mechanics. Until it is observed, Quantum Mechanics suggests that it exists in an ill-defined superposition of its possible states; in this case, 'decayed' and 'not decayed'. If this is true, the act of observation is clearly important in determining reality. The famous 'Schrödinger's Cat' paradox was formulated to highlight the apparent absurdity of this situation.

The concept of the collapse of a wavefunction can be seen as unrealistic in terms of Special Relativity. A wavefunction that describes two or more particles can extend over large distances in space, so there is no way that it can collapse 'instantaneously'. So, two photons produced by the decay of, say, a pion, will shoot off in opposite directions with equal but opposite momentum and spin. As far as Quantum Physics is concerned, the photons exist as superpositions of the possible states of spin, position and momentum. But it is possible to observe the spin and the position of one of the photons, even though it may be very far away from the other. According to Quantum Physics, at the exact moment you observe one photon, the entire wavefunction must collapse, forcing the other photon to take on a definite state of spin and position.

But this implies that the photons are somehow in ghostly, faster-than-light communication with each other. One way out of this apparent paradox surrounding the superposition of quantum states is to assume that quantum systems are in fact well-defined: that they have 'hidden variables'. So, for example, it makes common sense to assume that each photon actually has well defined properties all the time, and that Quantum Physics is incomplete. This was the position taken by Einstein, along with Boris Podolsky (1896 - 1966) and American-born Israeli physicist Nathan Rosen (1909 - 1995). In 1935, they set out to highlight the absurdity of the standard interpretation of quantum physics (the Copenhagen Interpretation). The result was the 'Einstein-Podolsky-Rosen' (EPR) paradox.

Several explanations have been put forward to make sense of the strange features of the quantum world, including the idea that rather than existing as a superposition of all possible states in a single Universe, a particle exists in one state in each of a number of different, 'parallel' Universes. This is the Many Worlds Interpretation of Quantum Physics – and, strange though it seems, it is one of the most elegant solutions to the strangeness of quantum behaviour. It might just be that there are an infinite number of universes, all differing very slightly, and created on the fly as wavefunctions collapse.

What do you know? You are trapped inside your mind. You think, therefore you are: your existence is all you can know for certain. You may well be imagining this book, your surroundings, even your body and your memories. You can never be sure. Subconsciously, you construct your perception of reality based, you assume, on sensory information from physical reality 'out there'. It could be that a supercomputer is feeding your brain a precise virtual reality model of a Universe; alternatively, your subconscious mind may be fabricating a fiction. In either case, what you think of as reality bears no resemblance whatsoever to what the world is really like. It is impossible to know what reality is, or even if any objective reality exists at all.

ABOUT THE AUTHOR

Jack Challoner has been a science writer for more than 25 years, and is the author of more than 40 books on science and technology. He has been shortlisted for many book prizes, including the Royal Society Book Prize and the Royal Society of Biology Book Prize in the UK, and the American Association for the Advancement of Science 'Excellence in Science Books' Prize in the USA. He has written on almost every aspect of science, but his first degree was in physics, so this book is very close to his heart.

Manufactured by Amazon.ca
Acheson, AB

11557380R00097